"十三五"普通高等教育本科部委级规划教材

画出20世纪服装风格（双语版）

DRAWING THE FASHION STYLES IN 20TH CENTURY

赵丽妍　李艳梅　刘慧艳　|　编著

中国纺织出版社有限公司
国家一级出版社
全国百佳图书出版单位

内 容 提 要

本书为"十三五"普通高等教育本科部委级规划教材。

夏奈尔曾经说过:"流行稍纵即逝,风格永存"。从古罗马风格开始,历史上已经出现了几十种服装风格,有些风格随着时间推移再次流行起来,而有些风格却再也没有出现过。

本书主要介绍20世纪百年间出现的服装风格,以十年为一个周期,图文并茂详尽介绍这个时间段服装风格的特征。

本书最大的特色是所有的图片均为服装效果图,图片根据对应年代的真实服装绘制而成,可以说既是一本服装风格书,也是一本服装绘画书,同时中英文对照,可以为越来越多的留学生提供学习资源。

本书适合高等院校的服装专业师生、留学生及服装设计师和服装绘画爱好者学习使用。

图书在版编目(CIP)数据

画出20世纪服装风格 = Drawing the Fashion Styles in 20th Century:汉英对照 / 赵丽妍,李艳梅,刘慧艳编著. — 北京:中国纺织出版社有限公司,2019.12

"十三五"普通高等教育本科部委级规划教材
ISBN 978-7-5180-6794-7

Ⅰ.①画… Ⅱ.①赵…②李…③刘… Ⅲ.①服装—绘画技法—双语教学—高等学校—教材—汉、英 Ⅳ.①TS941.28

中国版本图书馆 CIP 数据核字(2019)第 217485 号

策划编辑:魏　萌　　特约编辑:籍　博
责任校对:寇晨晨　　责任印制:王艳丽

中国纺织出版社有限公司出版发行
地址:北京市朝阳区百子湾东里 A407 号楼　邮政编码:100124
销售电话:010—67004422　传真:010—87155801
http://www.c-textilep.com
中国纺织出版社天猫旗舰店
官方微博 http://weibo.com/2119887771
北京华联印刷有限公司印刷　各地新华书店经销
2019 年 12 月第 1 版第 1 次印刷
开本:787×1092　1/16　印张:10
字数:126 千字　定价:58.00 元

凡购本书,如有缺页、倒页、脱页,由本社图书营销中心调换

前言 PREFACE

　　服装风格指一个时代、一个民族、一个流派或一个人的服装在形式和内容方面所显示出来的价值取向、内在品格和艺术特色。服装风格表现了设计师独特的创作思想，艺术追求，也反映了鲜明的时代特色，服装设计所追求的境界本质是风格的定位和设计。服装风格所反映的客观内容主要包括：时代特色、社会面貌及民族传统；材料、技术的最新特点和它们审美的可能性；服装的功能性与艺术性的结合。服装款式千变万化，形成了许多不同的风格，有的具有历史渊源、有的具有地域渊源、有的具有文化渊源，适合不同的穿着场所、不同的穿着群体、不同的穿着方式，展现出不同的个性魅力。

　　关于书名，我琢磨了很久。之前一直想编写一本有关时装绘画的书，但后来发现纯粹时装绘画类的书籍很多，而关于服装风格类的书却不是很多。与千变万化的服装款式不同，服装风格是相对固定的，于是我萌生出一个大胆的想法：将时装绘画与服装风格结合在一起，即书中的图片全部是时装绘画。当然，这些时装绘画是严格根据对应风格的真实图片绘制而成，将编者对时装绘画的浓厚感情投入到服装风格的介绍中，服装风格与服装时装绘画相得益彰。

　　随着我国高等教育国际化程度的不断提高，来华留学的人数稳步增加，这对高等学校的留学生教育提出了更高的要求，但与之相适应的教材仍比较匮乏。迄今为止，编者已多次为留学生开设两门全英文课程《时装绘画》和《时装设计》，在教学实践中，尝试用手绘的方式来表现服装风格，受到留学生们的欢迎。通过对教案的不断修改完善，以此为基础形成双语（中英文）讲义，值此机会，将讲义整理付诸出版，可以为留学生教育提供更好的教学资源。

这本书对服装专业人士来说，既是一本风格书，更是一本时装手绘书。本书可以供服装设计专业的国内在校生、留学生，服装设计及时装绘画的业余爱好者使用。全书共分为三大章：第一章主要是介绍服装风格的背景和分类；第二章着重笔墨介绍20世纪的服装风格；第三章主要是对不同风格所对应的服装设计要素以及服饰品时装画表现手法进行赏析。另外，本书包含大量图片以及针对性较强的思考练习题，对学生的专业学习和设计思维开拓具有一定的指导和启发作用。

本书的中文部分由赵丽妍负责，大部分画稿由赵丽妍和刘慧艳完成，英文部分由李艳梅负责。值此机会，我要感谢为本书奉献部分画稿的学生，他们是阮也一和尤心如。

同时感谢家人的大力支持和鼓励！

编著者

2019年3月

教学内容及课时安排

章/课时	课程性质/课时	节	课程内容
第一章/4	背景介绍/4	·	**绪论**
第二章/10	基础理论/30	·	**20世纪早期的服装风格**
		一	20世纪初至20年代服装风格
		二	20世纪30年代服装风格
第三章/10		·	**20世纪中期的服装风格**
		一	20世纪40年代服装风格
		二	20世纪50年代服装风格
		三	20世纪60年代服装风格
第四章/10		·	**20世纪后期的服装风格**
		一	20世纪70年代服装风格
		二	20世纪80~90年代服装风格
第五章/14	应用实践/14	·	**服装风格的时装画表现及欣赏**

注 各院校可根据自身的教学特色和教学计划对课程时数进行调整。

| 背景介绍 | 第一章 | 绪论 / 002 |
| | | Introduction |

基础理论	第二章	**20世纪早期的服装风格 / 026**
		Fashion Styles in the Early 20th Century
	第一节	20世纪初至20年代服装风格 / 026
		Fashion Styles From the Early 20th Century to the 1920s
	第二节	20世纪30年代服装风格 / 047
		Fashion Styles in 1930s

	第三章	**20世纪中期的服装风格 / 058**
		Fashion Styles in Mid-20th Century
	第一节	20世纪40年代服装风格 / 058
		Fashion Styles In 1940s
	第二节	20世纪50年代服装风格 / 060
		Fashion Styles in 1950s
	第三节	20世纪60年代服装风格 / 067
		Fashion Styles in 1960s

	第四章	**20世纪后期的服装风格 / 094**
		Fashion Styles in the Late 20th Century
	第一节	20世纪70年代服装风格 / 094
		Fashion Styles in 1970s

第二节　20世纪80~90年代服装风格 / 107

Fashion Styles From the 1980s to the 1990s

应用实践

第五章　服装风格的时装画表现及欣赏 / 122

Performance and Appreciation of Fashion Drawing

案例一　波西米亚风格 / 123

Example 1　Bohemian Style

案例二　波普风格 / 130

Example 2　Pop Style

案例三　解构风格 / 134

Example 3　Deconstruction Style

案例四　雅皮风格 / 139

Example 4　Yuppie Style

案例五　1920年代初期服装风格 / 144

Example 5　Fashion Styles in the Early 1920s

参考文献 / 151

Reference

背景介绍

课题名称：绪论
Project name: Introduction

课程内容：服装风格的概念，服装风格的分类
Course content: Concept of Fashion Styles, Classification of Fashion Styles

课题时间：4学时
Project time: 4 hours

训练目的：详尽阐述服装风格的概念和分类，使学生充分了解从古至今服装中出现的各类风格，为接下去的设计奠定基础。
The purpose of training: Elaborate the concept and classification of fashion styles; fully understand the various styles appearing in the fashion from ancient times to the present; lay the foundation for the next design.

教学要求：1. 使学生了解服装风格的概念和内涵。
　　　　　2. 使学生了解历史上不同时期服装风格分类的依据和特点。
　　　　　3. 使学生了解20世纪前后服装风格的显著区别。
Teaching requirements:
　　1. Make students understand the concept and connotation of fashion style.
　　2. Make students understand the basis and characteristics of fashion style classification in different periods in history.
　　3. Make students aware of the significant differences in fashion styles around the 20th century.

课前准备：阅读相关服装史方面的书籍。
Pre-class preparation: Read books on the history of fashion.

课后练习：归纳总结从古至今不同服装风格中的服装廓形的变化。
After-school exercises: Summarize the changes of the silhouette in different fashion styles from ancient times to the present.

第一章 绪论
Introduction

服装风格指一个时代、一个民族、一个流派或一个人的服装在形式和内容方面所显示出来的价值取向、内在品格和艺术特色。服装设计追求的境界说到底是风格的定位和设计，服装风格表现了设计师独特的创作思想，艺术追求，也反映了鲜明的时代特色。

服装风格所反映的客观内容，主要包括三个方面，一是时代特色、社会面貌及民族传统；二是材料、技术的最新特点和它们审美的可能性；三是服装的功能性与艺术性的结合。服装风格应该反映社会的时代面貌，在一个时代的潮流下，设计师们各有独特的创作天地，能够创造出百花齐放的繁荣局面。

如今，服装款式千变万化，形成了许多不同的风格，有的具有历史渊源、有的具有地域渊源、有的具有文化渊源，以适合不同的穿着场所、不同的穿着群体、不同的穿着方式，展现出不同的个性魅力。

根据时代划分，服装风格可以分成以下几大类，在每个时间段，又细分出不同的风格。

Fashion style refers to the value orientation, inner character and artistic characteristics of the fashion and the appearance of a certain era, a nation, a genre or a person. The realm of fashion design is the positioning and design of style. The style of fashion expresses the designer's unique creative thinking, artistic pursuit, and also reflects the distinctive characteristics of the times.

The objective content reflected in the style of fashion mainly includes three aspects. One is the characteristics of the times, the social appearance and the national tradition; the second is the latest features of materials and technology and their aesthetic possibilities; the third is the combination of functional and artistic aspects of fashion. The style of fashion should reflect the social outlook of the times. Under the trend of an era, designers have their own unique creations, which can create a flourishing prosperity.

Nowadays, the styles of fashion are ever-changing, and many different styles have been formed. Some have historical origins, some have local origins, and some have cultural origins. These are suitable for different wearing places, different wearing groups, different wearing methods and different personality charm.

According to the era, fashion styles can be divided into the following categories, and in each period, different styles are subdivided.

一、古罗马风格

古罗马风格的服装具有宽松、肥大、披挂式、缠绕式、多褶皱等特点。古罗马女性主要穿"斯朵拉"和"帕拉",男性穿"托嘎"。无论男女,在外袍的里面都会穿一种叫作"基同"的筒状衣服(图1–1)。

Ancient Rome Style

Ancient Rome Style is mainly characterized by loose, large, draped, wrap, and wrinkled. Ancient Roman women mainly wear "Stola" and "Palla", and men wear "Toga". Both men and women wear a tubular dress called "Chiton" in the robe (Figure 1-1).

图1-1 古罗马风格
Figure 1-1 Ancient Rome Style

二、哥特风格

受建筑影响，哥特风格的服装修长苗条，优雅精致且复杂华丽。注重腰部的设计，上身合体，裙子从臀部开始慢慢展开。"苏考特"是当时很流行的外衣（图1–2）。

三、巴洛克风格

巴洛克一词源于葡萄牙语，本意是有瑕疵的珍珠，引申为畸形的、不合常规的事物，在艺术史上却代表一种风格。巴洛克艺术在形式上表现出怪异与荒诞，豪华与矫饰的现象。巴洛克时期的服饰具有虚华矫饰的风格，极尽夸张雕琢之能事。巴洛克时期的法国服装非常奢华，常采用花缎、天鹅绒和锦缎等华贵面料，并带有大量镶嵌线和刺绣。紧身胸衣和裙撑将女性的曲线勾勒得淋漓尽致（图1–3）。

四、新古典主义风格

新古典主义风格兴起于18世纪的中期，其精神是巴洛克风格与洛可可风格之间产生的一种强烈反叛。它主要是力求恢复古希腊和古罗马所强烈追求的庄重与宁静感之题材与形式。特别是在女装方面。例如，以自然简单的款式，取代华丽而夸张的服装款式；又如，排除受约束、非自然的裙撑架等。因此在1790~1820年间，追寻淡雅、自然之美，在服装史上被称为

Gothic Style

Influenced by the architecture, Gothic Style fashion is long and slender, elegant and sophisticated. Focus on the waist design, the upper body fits, the skirt spreads slowly from the hips. "Surcoat" was a popular coat at the time (Figure 1-2).

Baroque Style

The word baroque is derived from the Portuguese which is intended to be a flawed pearl, a deformed and unconventional thing, but represents a style in art history. Baroque art shows strange and absurd, luxurious and pretentious phenomena in form. The style of the Baroque period has a style of illusion, and it is extremely exaggerated. The French style of the Baroque period is very luxurious, often with luxurious fabrics such as satin, velvet and brocade, with a large number of inlaid lines and embroidery. The bodice and skirt showed the curves of the woman vividly (Figure 1-3).

Neoclassicism Style

Neoclassicism style emerged in the mid-18th century, and its spirit is a strong rebellion against Baroque Style and Rococo Style. It is mainly aimed at restoring the theme and form of the dignity and tranquility that the Ancient Greek and Romans strongly pursued. Especially in women's fashion. For example, replace the gorgeous and exaggerated fashion style with a natural and simple style; exclude constrained, unnatural skirts, and so on. Therefore, from 1790 to 1820, the elegant and natural beauty pursued is

第一章　绪论
Introduction

图1-2　哥特风格的服装
Figure 1-2 Gothic Style Garment

图1-3　巴洛克风格的服装
Figure 1-3 Baroque Style Garment

新古典主义风格（图1-4）。

五、浪漫主义风格

浪漫主义风格服装主要强调女性化风格。它代表了一种强调自发性和视觉表现并凸现直觉和想象力的观点。可以用个性的、变化丰富的、出乎意料的、没有穷尽的、未完成的、过分的、解体的、抒情的和热烈奔放的等词汇来加以形容。装饰手段多用毛边、流苏、刺绣、花边、抽褶、蝴蝶结、花饰等，也就是说只要能想到的新鲜华丽的元素都可以被采用（图1-5）。

六、新艺术风格

从1890年起，西方女装进入了S型时期。西方女装紧身胸衣在前面把胸高高托起，将腹部压平，腰勒细，后面紧贴背部，把丰满的臀部自然地表现出来，从腰向下摆，裙子像小号似的自然张开，形成喇叭状波浪裙，从侧面观察时，宛如"S"型，因此而得名（图1-6）。

七、1920年代风格

第一次世界大战之后，女装发生了革命性的变化。女权运动是其中最重要的影响因素之一。一批新潮的、职业化的女性的涌现，促使职业装应运而生。人们不再需要那种使身体扭曲变

called Neoclassicism Style in the history of fashion (Figure 1-4).

Romanticism Style

Romanticism Style basically appears to emphasize the feminine style. It represents a view that emphasizes spontaneous and visual performance as well as highlighting intuition and imagination. It can be described in terms of individuality, rich and varied, unexpected, unfinished, excessive, disintegrated, lyrical, and enthusiastic. The decorative means mostly use raw edges, tassels, embroidery, lace, skirts, bows, floral ornaments, etc, that is, as long as the fresh and gorgeous elements that can be thought of could be used (Figure 1-5).

Art Nouveau Style

Since 1890, western women's wear has entered the S-shaped period. Women's corsets hold the front chest high, the abdomen is flattened, the waist is thin, close to the back, the full hips are naturally expressed, and from the waist to hem, and the skirt opens like a trumpet, forming flare skirt, when viewed from the side, the shape is like the "S", hence the name (Figure 1-6).

Fashion Styles in 1920s

After the First World War, women's fashion had undergone revolutionary changes. The feminist movement was one of the most important factors. The emergence of some modern and professional women

形的紧身衣，开始拒绝那些强调女性特征的设计，而是需要更多的腿部自由。由此便出现了简单宽松的直筒连衣裙和直筒短裙。另一方面，晚装则采用明亮色彩的闪光面料，并带有很多女性化的细节，如宽摆裙、褶裥裙，以及珍珠、亮片、流苏等装饰。夏奈尔可以说是20世纪20年代风格的化身（图1-7）。

had prompted the emergence of professional fashion. People didn't need the kind of tights that distort the body, and began to reject designs that emphasize feminine features, but need more leg freedom. This resulted in a simple loose straight dress and a straight skirt. On the other hand, the evening wear featured a brightly colored glitter fabric with many feminine details such as a wide swing skirt, pleated skirt, and pearls, sequins, tassels and more. Chanel could be said to be the embodiment of the 1920s style (Figure 1-7).

图1-4 新古典主义风格的服装
Figure 1-4 Neoclassicism Style Garment

画出 20 世纪服装风格 Drawing the Fashion Styles in 20th Century

图 1-5 浪漫主义风格的服装
Figure 1-5 Romanticism Style Garment

第一章　绪论
Introduction

图1-6　新艺术风格的服装
Figure 1-6 Art Nouveau Style Garment

画出 20 世纪服装风格 Drawing the Fashion Styles in 20th Century

图 1-7　1920 年代服装风格
Figure 1-7 Fashion Styles in 1920s

八、1930年代风格

1930年代是以经济大危机开始，以第二次世界大战开始而结束的。人们沉迷奢华，厌倦了模仿男孩子的女装风格，转而追求更加具有女人味的时装。因此，在1930年代中期以后，在欧美出现了一个追求典雅、苗条的时装阶段。从1930年代开始，裙子开始往下降，长裙子时代回归，苗条的腰身开始复兴。与以往不同的是，20世纪30年代的服饰是从古希腊服饰中寻求的设计灵感，富有优美的悬垂感（图1-8）。

九、1940年代风格

第二次世界大战（1939~1945年）期间的物资短缺直接影响着服装业。与其他很多日用品一样，服装也是限量供应。在这种情况下，服装的款式都变得又短又小。女装裙子的褶裥数量受到限制，袖子、领子和腰带的宽度也有相应的规定。刺绣、毛皮和皮革的装饰都受到禁止。裙长及膝而且裁剪得很窄。套装的设计注重功能性，并且适合各种场合穿着。其款式常常与军服相似，给人留下印象最深的是宽宽的垫肩和系得紧紧的腰带。而到1947年，高级时装在巴黎复苏，巴黎再度引领时尚的潮流。经历了多年的经济萧条之后，人们渴望穿上漂亮的服装，而迪奥意识到了这一点。他的时装发布引起的势不可挡的轰

Fashion Styles in 1930s

The 1930s began with a major economic crisis and ended with the beginning of the Second World War. People were addicted to luxury, tired of imitating the style of boy's fashion, and instead pursue fashions that were more feminine. Therefore, after the mid-1930s, there was a fashion stage that pursued elegance and slimness in Europe and America. Since the 1930s, the skirts had begun to fall, the long skirts had returned, and the slim waist had been resurrected. Different from the past, the fashion of the 1930s sought inspiration from Ancient Greek costumes, with a beautiful drapability (Figure 1-8).

Fashion Styles in 1940s

The shortage of supplies during the Second World War (1939~1945) directly affected the fashion industry. Like many other daily necessities, fashion was also available in limited quantities. In this way, the style of the fashion had become short and small. The number of pleats in women's skirts was limited, and the width of the sleeves, collars and belts were also specified. Embroidery, fur and leather decorations were prohibited. The skirt was long to knees and very narrow. The design of the suit was functional and suitable for all occasions. Its style was often similar to military uniforms, and the deepest impression was the wide shoulder pads and tight belts. By 1947, haute couture had recovered in Paris, and Paris once again led the fashion trend. After years of economic depression, people were eager to wear beautiful clothes, and Dior realized this. The unstoppable sensation caused by his shows how

画出 20 世纪服装风格 Drawing the Fashion Styles in 20th Century

图 1-8　1930 年代服装风格
Figure 1-8 Fashion Styles in 1930s

动,表明迎合人们的这种渴望是多么成功。随着充满女性化的"新式样"的推出,迪奥一举成为"时装之王"。这种款式的特点是:长及小腿,裙子下摆宽大,上衣肩部圆润,充分展现了女性优美的形体。另外一种年轻有活力的风格是优雅女性化的铅笔形细长风格,这种风格的特点是上衣部分紧身合体,裙子修长瘦窄(图1–9)。

十、1950年代风格

随着高级时装向可穿性时装转变,使得社会各阶层人都得以享用,服装业发生了巨大的变化。人们的社会地位可以通过服装加以强调。每个季节都有新的女装流行。强调女性性别特征的时装以线条修长、收腰和臀部修饰为特点。裙摆从小腿上移至膝盖。外形轮廓均用字母或形状命名。典型的例子包括铅笔型、郁金香型、Y型。年轻时髦的例子包括带有公主线的X型、带有裙撑和内裙的圆顶屋型和气球型以及阔摆的梯型轮廓和A型。H型和喇叭型的特征是拥有腰线流畅的衬衫式上衣(图1–10)。

十一、1960年代风格

20世纪60年代服装特征是冲破传统的限制和禁忌,广告和媒体中最流行的词汇是年轻。玛丽·匡特的迷你裙是当时最典型的流行风格。运动且休闲的宽松学生裙和衬衫裙很流行。

successful it was to cater to people's desires. With the launch of the feminine "New Style", Dior became the "King of Fashion". The characteristics of this style were: long to calf, wide bottom, round shoulders, fully showing the beautiful shape of women. Another young and energetic style was the elegant feminine pencil-shaped slender style, which was characterized by a tight top and a slim skirt (Figure 1-9).

Fashion Styles in 1950s

With the transformation of high fashion into wearable fashion, people of all levels of the ceremonies had been able to enjoy, and the fashion industry had undergone tremendous changes. People's social status could be emphasized through fashion. New women's fashion was popular every season. Fashions that emphasize the gender characteristics of women were characterized by slenderness, waist and hips. The hem of skirt moved from the calf to the knee. The outlines were named with letters or shapes. Typical examples included pencil, tulip, Y-shaped. Young fashionable examples of included the X-shaped with princess line, the dome and balloon with bustles and inner skirts, and the trapezoidal shape and A-shaped with wide swing. The H-shaped and flare shape were characterized by a blouse with smooth waistline (Figure1-10).

Fashion Styles in 1960s

In the 1960s, fashion was characterized by breaking through traditional restrictions and taboos. The most popular vocabulary of advertising and media was young. Mary Quant's mini skirt was the most typical

画出 20 世纪服装风格 Drawing the Fashion Styles in 20th Century

图 1-9　1940 年代服装风格
Figure 1-9 Fashion Styles in 1940s

第一章 绪论
Introduction

图1-10

画出 20 世纪服装风格 Drawing the Fashion Styles in 20th Century

图 1-10　1950 年代服装风格
Figure 1-10　Fashion Styles in 1950s

衬衫、长马甲和无领无袖连衣裙也很流行。夏奈尔套装和女裤开始被人们接受并逐渐成为经典。太空旅行和抽象派艺术带来了以黑色与白色、白色与银色为特征的几何图形和未来主义风格。反主流的嬉皮士风格也影响着服装界。（图1-11）。

pop style at the time. Sports and leisure loose student skirts and shirt skirts were very popular. Shirts, long vests and collarless sleeveless dresses were also popular. Chanel suits and trousers began to be accepted and become classics. Space travel and abstract art bring geometry and futuristic styles characterized by black and white, white and silver. The anti-mainstream Hippies Style also affected the fashion industry (Figure 1-11).

图1-11 1960年代服装风格
Figure 1-11 Fashion Styles in 1960s

十二、1970年代风格

20世纪70年代服装适应范围很宽，可以让人们组合自己独特的风格。流行将单件购买的服装进行组合，同样也流行面料的混合及板型的混搭。裙长有些波动，最终定在中等长度。日装的混搭主要包括褶裥迷你裙配衬衫式连衣裙、喇叭裤、宽松衬衫配短外套。在混搭流行期间，出现了很多款型各异的女衫、裤子和外套。裙摆降到了膝盖以下。传统而浪漫的风格成为新的流行趋势。新的性感款式采用有悬垂感的面料，上衣部分很合体，裙子较长，用腰带强调腰部，并采用褶边、荷叶边及刺绣来突出这一风格。异国情调对晚装的影响很大。最后出现了工装风格的服装，这种款式直线裁剪、强调肩部。超号型风格的出现清楚地表明流行走向更加休闲化的趋势。在1970年代还出现了波西米亚风格、朋克风格、中性风格等（图1-12）。

十三、1980年代风格

1980年代风格突出了女装的职业化。20世纪80年代是职业女性不断涌现的时代，女装呈现出向男性化靠拢的迹象，在服装结构、造型和细节上的表现尤其强烈。三件套套装（上衣、裤子或裙子、衬衫）是1980年代的产物，这种源自男装的着装形式本身体现出浓浓的女装男性化

Fashion Styles in 1970s

In 1970s, fashion had a wide range of adaptations, allowing people to combine their own unique styles. It was popular to combine garments purchased in a single piece, as well as the mixing of popular fabrics and the mixing of styles. The length of the skirt was somewhat fluctuating and was finally set at medium length. The combination of daywear mainly included pleated mini skirts with shirt dresses, flared pants, loose shirts with short jackets. During the combination of fashion, there were many different types of blouses, pants and jackets. The hem of skirt fell below the knee. Traditional and romantic styles had become a new trend. The new feminine style used drape fabric, the tops were well-fitted, the skirts were longer, the waist was accentuated with a belt, and the pleats, ruffles and embroidery accentuated the style. Exoticism had a great impact on evening wear. Finally, Labor style of fashion appeared, which was cut in a straight line and emphasized the shoulders. The emergence of the super-type style clearly showed that the trend was more casual. In the 1970s, there were also Bohemian Style, Punk Style, and Neutral Style and so on (Figure 1-12) .

Fashion Styles in 1980s

Styles in 1980s highlights the professionalism of women's wear. The 1980s was an era in which professional women continued to emerge. Women's wear showed signs of masculinity, especially in the structure, shape and details of fashion. The three-piece suit (tops, trousers or skirts, shirts) was the product of the 1980s. The dress form derived from men's wear itself reflected the tendency of masculine masculinity. The

第一章 绪论
Introduction

图1-12

第一章　绪论
Introduction

图 1-12　1970 年代的波西米亚风格
Figure 1-12　Bohemian Style in 1970s

倾向。1980年代的总体特征是大,甚至是巨大,外轮廓造型,款式细节,甚至服饰配件都呈现宽大特征,这也是1980年代和其他风格的主要区别。在这个时期,出现了雅皮风格、解构风格等风格(图1-13)。

overall characteristics of the 1980s were large, even huge, contours, fashion details, and even accessories showed a wide range of features, which was the main difference between the 1980s and other styles. During this period, styles such as Yuppie Style and Deconstruction Style appeared (Figure 1-13).

图1-13 1980年代服装风格

Figure 1-13 Fashion Styles in 1980s

本章小结 Conclusions

1. 1950年代之后，服装风格变得多样化。

    More and more fashion styles appeared after 1950s.

2. 1960年代的服装摆脱了各种禁忌。

    Fashion in 1960s got rid of many taboos.

思考题 Thinking questions

1. 1960年代的口号是年轻化，导致这种年轻文化流行的原因是什么？

    What made youth as the slogan of 1960s?

2. 1970年代出现的波西米亚风格的特点是什么？

    What are the characteristics of Bohemian styles that emerged in the 1970s?

基础理论

课题名称：20世纪早期的服装风格
Project name: Fashion Styles in the Early 20th Century

课程内容：20世纪初至20年代的服装风格，20世纪30年代服装风格
Course content: Fashion Styles From the Early 20th Century to the 1920s, Fashion Styles in 1930s

课题时间：10学时
Project time: 10 hours

训练目的：详尽阐述服装从20世纪初至30年代末的风格变化。
The purpose of training: Explain the fashion styles changed from the beginning of the 20th century to the end of the 1930s.

教学要求：1. 使学生了解1930年代初多种风格并存的现象和原因。
　　　　　2. 使学生了解战争给服装带来的影响。
　　　　　3. 使学生了解1930年代服装长度发生的翻天覆地的变化。
Teaching requirements:
　　　1. Make students understand the phenomena and causes of the coexistence of various styles in the 1930s.
　　　2. Make students understand the impact of war on fashion.
　　　3. Make students understand the earth-shaking changes in the length of fashion in the 1930s.

课前准备：阅读相关服装美学和服装史方面的书籍。
Pre-class preparation: Read books on the history and fashion aesthetics of fashion.

课后练习：将20世纪早期的不同服装风格整理出来，然后分析它们之间的不同点。
After-school exercises: Sort out the different styles of fashion in the early 20th century and analyze the differences between them.

第二章 20世纪早期的服装风格
Fashion Styles in the Early 20th Century

第一节 20世纪初至20年代服装风格
Fashion Styles From the Early 20th Century to the 1920s

在现代服装设计出现之前，欧美上层社会的妇女也穿着讲究，她们的服装是由讲究的裁缝精心制作的，而这些裁缝并不在服装上署名，也没有自己的品牌或服装店，他们是传统的匠人，法语称为"fournisseur"，也就等于中文的裁缝。服装的形式基本一样，都是称为裙衫的女装，包括紧凑的上身部分，宽大的裙子，强调胸部，臀部也突出，小腹平直，衣领高耸，加上夸张的帽子和帽子上复杂庞大的鸵鸟毛装饰，形成所谓的S型服装样式。由于上身极为紧凑，而下部的裙子则宽大拖沓，因此也称为A型服装，整个设计的核心内涵，就在于紧贴身体，必须有紧身胸衣或者紧身马甲，胸衣把女性的身体都束缚成为一个标准的式样。在当时，服装的设计不是要达到个性特征，而恰恰相反，是要使女性在穿着上显得一样。模式化和标准化，是当时整个社会崇尚的方式。

这种S型的浪漫风格与当时追求自由和服装改革的自由派女性的期望相违背，她们希望摆脱传统，摆脱束缚

Before the advent of modern fashion design, women in the upper echelons of Europe and the United States also wore exquisite clothes. Their clothes were carefully crafted by elaborate tailors. These tailors were not signed on fashion, nor did they have their own brands or fashion stores. They were traditional craftsmen, called "fournisseur" in French, which was equal to the tailor of Chinese. The fashion was basically in the same form, and was called a crinoline dress, including a compact upper body, a large skirt, emphasizing the chest, buttocks, straight belly, tall collar, plus exaggerated hat and complex and huge ostrich feather. All of these formed a so-called S-shaped fashion style. Because the upper body was extremely compact, and the lower skirt was wide and dragged, it was also called A-shaped fashion. The core connotation of the whole design was close to the body. There must be a corset or a tight vest, and the corset binds the female body into a standard style. At that time, the design of fashion was not to achieve personality characteristics, on the contrary, it was to make women appear the same in appearance. Modularity and standardization was the way that the whole society advocated at that time.

The Romantic Style of this S-shaped was contrary to the expectations of liberal women who pursued

第二章　20世纪早期的服装风格
Fashion Styles in the Early 20th Century

她们身体的紧身胸衣和紧身马甲。于是，保罗·波列，一名真正的服装设计师，以摧枯拉朽的能力推翻了紧身胸衣控制服装的长期垄断，创造了新的服装。保罗·波列从古希腊、古罗马和东方服装上寻找灵感，因此这个时期的服装带有明显的古典风格和东方风格。同时受到当时流行的"新艺术运动"的影响，也受到18世纪法国的"执政风格"的影响，他设计的服装呈现装饰艺术风格和帝政风格。保罗对当时流行的女权运动所主张的服装进行了研究，从而提出了自己服装设计的新方式，无论他推出什么风格的服装，都深得当时女性的喜爱。20世纪初期所流行的古希腊、古罗马风格服装，除了保罗·波列的设计，还有一位设计师就是玛丽亚诺·佛特尼，她最为著名的设计就是那件灵感来自古希腊"迪佛斯"的晚装。它的设计好像古希腊的束腰外衣"基同"一样，这件衣服从肩部一直垂到脚面，之间没有任何褶缝、垫、缝边装饰，可以说在设计上简直朴实无华。"迪佛斯"完全放弃了紧身胸衣，使穿它的女性得到活动的充分自由，身体可以自由舒展。

20世纪初期的巴黎是催生服装设计师的时期，也是风格变化较多的时期。代表设计师当属法国的可可·夏奈尔，她的设计以宽松内衣的基本结构加以演化而成，宽松、简

freedom and fashion reform at the time. They wanted to get rid of tradition and get rid of the corsets and tight-fitting vests which bound their bodies. So, Paul Poiret, a real fashion designer had overturned the long-term monopoly of tight-fitting bra-controlled garments with the ability to ruin and create new outfits. Paul Poiret looked for inspiration from Ancient Greek, Roman and Oriental costumes, so the costumes of this period had a distinct classical and oriental style. At the same time, he was influenced by the popular "New Art Movement", and was also influenced by the "Imperial Style" of France in the 18th century. Therefore, he designed the costumes to present the Art Deco style and the Imperial Style. Paul studied the fashion advocated by the popular feminist movement at that time, and proposed a new way of designing his own fashion. No matter what style of fashion he introduced, he was deeply loved by women at that time. About the Ancient Greek Rome Style popular in the early 20th century, in addition to the design of Paul Poiret, there was also a designer was Mariano Fortney, her most famous design was the inspiration from the Ancient Greek "Diffus" evening dress. It was designed like the Ancient Greek tuxedo "Chiton". This dress hangs from the shoulder to the foot, without any creases, cushions, seams, and it could be said that the design was simple and unpretentious. Diffus completely gave up the corset, so that the women wearing it were fully free to move and the body could stretch freely.

Paris in the early 20th century was a period of birth to fashion designers, and it was also a period of more style changes. The representative of the designer was the French Coco Chanel, her design evolved from the basic structure of loose underwear, loose, simple, and made of soft cotton fabric, in line with sports requirements. In this decade,

单并且采用松软的棉质面料，符合体育运动的要求。在这十年中，有一种风格对时装界影响较大。"俄罗斯芭蕾舞团"来巴黎的演出是一个革命化的冲击，这个芭蕾舞团令人耳目一新的剧目中，鲜艳、特殊的服装对于服装设计师来说是一个巨大的震动，也是一个重大的启发。这些演出服装的色彩绚丽、灿烂，打破了巴黎上层社会传统的阴暗、沉闷和保守且一成不变的服装形式，为巴黎服装设计带来了春天的气息。在俄罗斯芭蕾舞服装的刺激下，巴黎的服装也因此告别了装饰华贵、式样保守的旧时代，告别了"美好时光"时期的服装时代，服装设计进入具有创意的、新鲜的、年轻的、简单和朴素的、自然而生动的新阶段。

1918年，在第一次世界大战结束后，欧洲和美国的人们都感到应该享受来之不易的胜利，享受胜利果实，穿华丽的衣服，化妆浓艳，想把战争夺取的时间抢回来。因此助长了物质主义的泛滥。战后的欧洲妇女在服装和打扮上花费了相当大比例的收入，她们的新口号是"为今天而活"。青年人讲究时髦、时尚和崇拜偶像，这是1920年代最突出的现象。女孩子认为消瘦才是时髦，因此，20年代女子都拼命节食，努力消瘦，直到变成瘦骨嶙峋，才被视为美观。这种类型的女孩子喜欢小男孩的形象，短发、着男装、消瘦（图2-1）。

there was a style that had a great influence on the fashion industry. The performance of the "Russian Ballet" to Paris was a revolutionary impact. This refreshing repertoire of the ballet, bright and special costumes were a huge shock and a major inspiration for the fashion designer. The colorful and splendid colors of these costumes broke the dark, dull and conservative style of the traditional Parisian upper class, bringing a springy atmosphere to Parisian fashion. Stimulated by the Russian ballet costumes, the Parisian costumes also bid farewell to the old times of decorative luxury and conservative style. They bid farewell to the "Good Times" period, and fashion design enters creative, fresh, young, simple and a simple, natural and vivid new stage.

In 1918, after the end of the First World War, people in Europe and the United States felt that they should enjoy the hard-won victory, enjoy the fruits of victory, wear gorgeous clothes, bright make-up, and try to take back the time of war. This had contributed to the proliferation of materialism. After the war, European women spent a considerable proportion of their income on fashion and dressing. Their new slogan was "Live For Today". Young people paid attention to fashion and admired their favorite idols, which was the most prominent phenomenon in the 1920s. Girls think thinness was fashionable. Therefore, in the 1920s, women were desperately dieting, trying to lose weight until skinny, which was considered beautiful. This type of girls were fond of the image of a little boy—short hair, male fashion, and thin (Figure2-1).

第二章　20世纪早期的服装风格
Fashion Styles in the Early 20th Century

图 2-1

029

画出 20 世纪服装风格 Drawing the Fashion Styles in 20th Century

保罗·波列的灵感来自东方的作品
Paul Poiret's design work inspired from orient

第二章 20世纪早期的服装风格
Fashion Styles in the Early 20th Century

受俄罗斯芭蕾舞团影响的服装样式
Garments influenced by the "Russian Ballet"

图 2-1

画出 20 世纪服装风格 Drawing the Fashion Styles in 20th Century

1920 年代的女性以瘦弱为美，被称为"假小子"
Women in the 1920s were thin, referred to as "Tomboys"

图 2-1　20 世纪初期服装风格
Figure2-1　Fashion Styles in the Early 20th Century

一、浪漫主义风格

浪漫主义风格是将浪漫主义的艺术精神应用在时装设计中的风格,在服装史上,巴洛克和洛可可服饰具有浪漫主义的特征。1825~1850年的欧洲女装属于典型的浪漫主义风格,这个时期被称为浪漫主义时期。浪漫主义风格的服装主要表现在非活动性的女装上,特征为宽肩、细腰和丰臀。上衣用泡袖、灯笼袖或羊腿袖来加宽肩部尺寸,紧身胸衣造成丰满的胸部和纤细的腰肢,与圆台型的撑裙共同塑造成X型的造型线条。面料多为轻而柔软的薄棉布,织纹较密的白麻布,光亮飘逸的绸粉红色,白色较常用,此外还有黄、浅紫和紫色。在现代时装设计中,浪漫主义风格主要反映为柔和圆转的线条,变化丰富的浅淡色调,轻柔飘逸的薄型面料,以及泡袖、花边、绲边、镶饰、刺绣、褶皱等。

20世纪初以沃斯为代表的浪漫主义风格服装有以下几个特征:直筒型的夹克,细高并饰以硬质蕾丝装饰的领子将头部高高托起;装饰丰富的帽子;沉重的鸵鸟毛装饰是当时最流行的样式,并且是身份地位的象征。裙子是臀围宽松的钟形及地长裙,在臀部后面堆以大量的褶皱和荷叶边。为了平衡这样的造型,于是高高耸起的帽子被故意固定在头发上并向前倾斜,和服装相配套的鞋子有尖头和巴洛克式的弧形后跟。丝质面罩和细长的手套确保女性外出时面部和手臂的皮肤不被看到,甚至即使穿着短袖的衣服,也要戴上长长的手套,这

Romanticism Style

Romanticism Style is that applies the artistic spirit of romanticism to fashion design. In the history of fashion, Baroque and Rococo costumes have romantic characteristics. European women's fashion between 1825 and 1850 was a typical Romanticism style, this period is called the Romantic period. Romanticism Style fashion was mainly expressed in inactive women's fashion, characterized by wide shoulders, thin waist and rich buttocks. The top was made up of pop sleeves, lantern sleeves or lamb sleeves to widen the shoulder size. Full chest and slender waist, round-frame bustle forms the X-shaped. The fabrics were mostly light and soft cotton cloth, white linen with dense texture, bright and flowing silk pink, white was more common, in addition to yellow, light purple and purple. In modern fashion design, the romantic style is mainly reflected in the soft and round lines, the rich and light shades, the soft and elegant thin fabric, as well as the sleeves, lace, piping, inlays, embroidery, flares and so on.

At the beginning of the 20th century, the Romanticism Style represented by Voss had the following characteristics: a straight-type jacket, a high-rise collar decorated with hard lace, holding the head high; a richly decorated hat; heavy ostrich feather decoration was the most popular style at the time and a symbol of status. The skirt was a loose bell-shaped and long skirt with a large hip and a ruffle on the back of the buttocks. In order to balance such a shape, the hat was deliberately fixed on the hair and tilted forward, and the matching shoes of the clothing had sharp pointed toe and baroque curved heels. Silk masks and slender gloves ensure that the skin of the face and arms was not visible when

样一来，上臂的皮肤就不会暴露。不难想象，在当时，哪怕是露出一点点皮肤都足以让男士疯狂了。日装的面料主要是麻、天鹅绒和毛，色彩以暗色或一些如淡粉红色、淡蓝色或紫红色等粉彩色为主，然后配以大量的装饰，像珠子、彩带、珠管、蝴蝶结、贴花和假花等（图2-2）。

the woman went out, even wearing short-sleeved clothes, so that the upper arm's skin was not exposed. It's not hard to imagine that at the time, even a little exposed skin was enough to make men crazy. The fabrics of the day were mainly linen, velvet and fur. The colors were mainly dark or some pastel colors such as light pink, light blue or purple, and then decorated with a lot of decorations, like beads, ribbons, beads, bows, decals and fake flowers (Figure2-2).

羽毛是帽子上必不可少的装饰
Feather was an essential decoration for hats

第二章　20世纪早期的服装风格
Fashion Styles in the Early 20th Century

图 2-2　浪漫主义风格
Figure 2-2　Romanticism Style

二、东方风格

提到20世纪初期的东方风格，我们就不得不谈保罗·波列，他在服装设计中采用了很多东方的风格，特别是日本、中国、印度和阿拉伯世界的服装特点和风格。这种借鉴加上他早期引用的古希腊、古罗马的风格，形成了非常突出的形式，对于传统的欧洲女装来说，自然具有很大的冲击力量。

当时在欧洲上流社会很流行俄罗斯的芭蕾演出。1909年，俄罗斯芭蕾舞团在巴黎演出了两个芭蕾舞剧，极为成功，服装和舞台做了革命性的改革，东方风格和现代艺术气息混合在一起，影响了整个巴黎时装界。受此风格的印象，保罗·波列设计了一系列具有东方风格的时装，包括采用腰带的长袖衣衫、日本式样的和服、东方宽大的女子长裤、阿拉伯风格的女子束腰外衣、面纱、穆斯林式样的头巾，这些设计都非常典雅和流畅，加上鲜艳的色彩图案，得到巴黎和欧洲其他国家上层女子的喜爱，不少巴黎女子追随他的这些设计。在这些东方风格服装的设计中，保罗·波列对面料也进行了改革。他采用华丽的装饰：色彩鲜艳的刺绣、锦缎、流苏、珍珠和罕见的羽毛都是他广泛使用的装饰品。当时巴黎人人都陶醉于东方艺术，对这种艺术方式自然也接受和欢迎（图2-3）。

Oriental Style

When it comes to the oriental style of the early 20th century, we have to talk about Paul Poiret, who used a lot of oriental styles in fashion design, especially the fashion features of Japanese Style, Chinese Style, India Style and the Arab world. This kind of reference, combined with the Ancient Greek and Rome Styles quoted by him, had formed a very prominent form. For traditional European women's wear, it naturally had a great impact.

At that time, the Russian ballet performance was very popular in the upper class in Europe. In 1909, the Russian Ballet performed two ballets in Paris, which was extremely successful. The revolutionary reforms of fashion and the stage, mixed with Oriental Style and modern art, influenced the entire Paris fashion industry. Impressed by this style, Paul Poiret designed a series of Oriental Styles fashion, including long-sleeved shirts with belts, kimonos in Japanese Style, and pantaloons in the east, Arabian Style women's tunic, veil, and turban of Muslim Style, these designs were very elegant and smooth, with bright color patterns, which were popular among women in Paris and other European countries. Many women in Paris followed his designs. In the design of these Oriental Style garments, Paul Poiret had also reformed the fabric. He used gorgeous decorations: colorful embroidery, brocade, tassels, pearls and rare feathers for decorations. At that time, everyone in Paris was enchanted by oriental art, which naturally accepted and welcomed (Figure2-3).

第二章　20世纪早期的服装风格
Fashion Styles in the Early 20th Century

灵感来自古希腊的服装
Inspired by Ancient Greek Style fashion

图 2-3

037

灵感来自东方风格的设计
Inspired from Oriental Style

第二章　20世纪早期的服装风格
Fashion Styles in the Early 20th Century

灵感来自和服的设计
Inspired from the kimono

图 2-3

图 2-3 东方风格
Figure2-3 Oriental Style

三、装饰艺术风格

装饰艺术是20世纪20~30年代流行于法国的主要的艺术风格。这种艺术风格深受埃及和非洲古老装饰风格、立体派、野兽派、表现主义以及欧洲纯粹派的影响，同时，受俄罗斯芭蕾舞团的影响，1920年代的装饰艺术风格具有东方情调。对1920年代服装的装饰艺术风格影响最为深远的设计师当属让·巴铎。他擅长具有民族风格的刺绣设计，特别是具有装饰艺术风格的形式，再加上强烈的色彩，使他的设计深受女性们的喜爱。他在艺术上的立体主义和设计上的装饰艺术运动两个潮流中寻找借鉴，有意识地采用两者强烈的色彩和线条，突出的几何图形，来设计自己的服装。他的设计具有强烈的现代感，很受当时一些前卫的女性喜爱（图2-4）。

四、爵士风格

第一次世界大战使人们变得惶恐不安，生怕眼前的一切转瞬即逝，叛逆、奢侈、混乱和探索是这个时期的特点。这十年被称为咆哮的20年代，或疯狂的时代。在这个短暂的时期里，人们如痴如狂，物欲横流，醉生梦死，纸醉金迷。同时，这个时期也是爵士乐和查尔斯顿的演出风靡一时的时代，是女孩子留着被称为"泡泡头"短发并涂着鲜红的嘴唇的时代，是自由恋爱和香烟、避孕和短裙的时代，这些离经叛道的现象是这十年的象征（图2-5）。

Art Deco Style

Art Deco was the main artistic style popular in France in the 1920s and 1930s. This artistic style was deeply influenced by the ancient Egyptian and African decorative styles, Cubism, Fauvism, Expressionism and European Pureism. At the same time, influenced by the Russian Ballet, the Art Deco Style of the 1920s had an oriental atmosphere. The most influential designer of the Art Deco Style of the 1920s was Jean Barthew. He specialized in ethnic-style embroidery designs, especially in the form of Art Deco Style, coupled with strong colors, which made his designs popular with women. He sought to draw on the two trends of the artistic Cubism and the Art Deco movement in design, consciously adopting the strong colors and lines of both, and highlighting the geometric figures to design their own fashion. His design had a strong modern feeling and was very popular with some avant-garde women at the time (Figure2-4).

Jazz Style

The First World War made people fearful and uneasy, fearing that everything in front of them was fleeting, and that rebellion, luxury, chaos and exploration were the hallmarks of this period. This decade is known as the roaring 1920s, or crazy era. In this short period of time, people were crazy, materialistic, and drunk. This was an era famous for the Charleston and jazz, bobbled hairstyles, red lips, free love, cigarettes, birth control, and short skirts. These deviant phenomena were the symbol of this decade (Figure2-5).

画出 20 世纪服装风格 Drawing the Fashion Styles in 20th Century

20 世纪初的刺绣面料
Embroidered fabrics in the early 20th century

图案装饰在 20 世纪初很受欢迎
Decoration pattern was very popular in the early 20th century

第二章　20世纪早期的服装风格
Fashion Styles in the Early 20th Century

线条装饰艺术
Line deco art

图 2-4　装饰风格
Figure2-4 Art Deco Style

画出 20 世纪服装风格 Drawing the Fashion Styles in 20th Century

1920 年代出现的爵士风格服装，套头针织衫成为主流
Jazz-style fashion in the 1920s, pullover sweaters were popular

第二章　20世纪早期的服装风格
Fashion Styles in the Early 20th Century

吊带短裙、珍珠项链、毛饰、短发以及烟斗，成了爵士女孩的标配
Sling short skirt, pearl necklace, furry, short hair and pipe, became the standard for jazz girls

图 2-5

045

画出 20 世纪服装风格 Drawing the Fashion Styles in 20th Century

女孩们穿闪光面料服装、戴长手套
Girls dressed in flash fabrics and wore long gloves

图 2-5　爵士风格
Figure2-5　Jazz Style

第二节　20世纪30年代服装风格
Fashion Styles in 1930s

从1920~1929年，西方社会经历了巨大变化，欧洲经济经过第一次世界大战之后的萧条，这十年开始逐步恢复，社会消费因此趋向繁荣和奢侈，但是，随着时间的推移，世界进入了经济大萧条时期。人们从1929年开始的经济大危机中认识到稳定的脆弱性和人生的短暂性，因此对于生活的品质和享乐有了更大的追求。在服装上，人们已经厌倦了矫揉造作的模仿男孩子的女性服装和缺乏女性风采的女装，转而追求更加女性化的时装。因此，在这十年中，欧美都出现了一个追求典雅、苗条风格的阶段。这个时期的服装，是以极为优雅的设计在服装设计史上而著称的。

20世纪30年代的女人在晚上总是穿丝绸长裙，只有这种最昂贵的面料，才能通过斜裁产生流线型的效果，既不暴露身体又显示身材。马德琳·维奥内特斜裁的杰出新技术被广泛使用。这一发明的独特之处在于，在弹性莱卡发明之前，斜裁就可以赋予面料弹性。这些闪闪发光的丝绸晚礼服不需要系紧，只要从头顶穿上或踩进去就可以了。低领口在当时非常流行，通常只是简单地搭配着一串摇晃的珍珠项链。

From 1920 to 1929, western society experienced tremendous changes. The European economy that began in this decade had gradually recovered after the depression of the First World War, and social consumption had thus become prosperous and extravagant. However, as time goes by, the world then entered the Great Depression. People had realized the fragility of stability and the shortness of life from the economic crisis that began in 1929, and therefore had a greater pursuit of the quality and enjoyment of life. In clothing, people were tired of the imitation of women's clothing that imitates boys and the lack of femininity, and instead pursue a more feminine fashion. Therefore, in this decade, Europe and the United States had a stage of pursuing elegance and slim style. The clothing of this period was known for its extremely elegant design in the history of fashion design.

The women of the 1930s always wore long dresses in the evening, and it was silk or nothing, as only this, the most expensive fabrics, fell on the diagonal in such a streamlined way, flattering the body without exposing it. Madeleine Vionnet's brilliant new technique of cutting the material on the bias was universally copied. The ingenuity of this lay in the way the cut gave the material elasticity, long before the invention of lycra. These evening dresses of shimmering silk satin needed no fastening, but could simply be pulled on over the head or stepped into. The low-cut necklines were very popular at that time, and they were often simply yet effectively emphasized with a single dangling string of pearls.

对于裸露的肩膀，最好的东西当然是皮毛，尤其是银狐，在肩膀上套上两层银狐皮被认为是特别时髦的。然而，最吸引人的是白色狐皮制成的整个斗篷。买不起的人退而求其次用一条天鹅绒的披肩或者鲜艳色彩的雪纺围巾。

女人知道如何充分利用经济衰退。那些买不起新裙子的人会把她们的旧裙加长。当时，超短裙摆在任何场合下都不再流行了，裙长跌落到大约小腿中部，并且所有短下摆都被用丝带、嵌片、额外的材料或皮毛巧妙地制成合适的长度。同时在领口或袖口上添加少量的毛皮给衣服增添了一丝丝奢华。

有钱的妇女白天也穿皮衣。波斯羔羊、阔尾羊、海狸和水獭都被制成了四分之三身长的外套，同时穿上了必备的公主裙。连衣裙由一整块面料裁剪，腰部用细腰带强调。翻领很宽，深V领口，至少在夏天是这样，里面还穿了一件衬衫。一个淑女的装束总是包括手套和帽子。20世纪30年代最合乎情理的时尚就是和最具异国风格的头饰搭配在一起（图2-6）。

The best thing for bare shoulders was of course, fur, especially silver fox, and laying two whole skins about one's shoulders was seen as particular chic. The height of glamour, however, was an entire cape made of white fox fur. Anyone who could not stretch to that went for a velvet cape or a brightly colored chiffon wrap.

Women knew how to make the best use of the downturn. Those who could not afford to buy new dresses would simply lengthen the ones that they had. At the time, very short hemlines were no longer in fashion in any case, but fell to about mid-calf, and anything shorter was artfully made into the correct length with ribbons, panels, and extra material, or fur. Using the very smallest pieces of fur at the neckline or as cuffs gave fashion a hint of luxury.

Wealthy women wore fur during the day, too. Persian lamb, broadtail, beaver, and otter were also made into three-quarter-length coats and worn with the obligatory princess dress. This was cut from a single piece of material, and the waist was emphasized with a narrow belt. The lapels were wide and the necklines plunging, at least in summer, and a blouse was worn underneath. A ladylike outfit always included gloves and a hat. The sensible fashions of the 1930s were paired with the most outlandish headpieces (Figure 2-6).

第二章　20世纪早期的服装风格
Fashion Styles in the Early 20th Century

图2-6

第二章　20世纪早期的服装风格
Fashion Styles in the Early 20th Century

图2-6

画出 20 世纪服装风格　Drawing the Fashion Styles in 20th Century

第二章　20 世纪早期的服装风格
Fashion Styles in the Early 20th Century

图 2-6

画出 20 世纪服装风格 Drawing the Fashion Styles in 20th Century

第二章　20世纪早期的服装风格
Fashion Styles in the Early 20th Century

图 2-6　女性的优雅气质在 1930 年代体现得淋漓尽致
Figure2-6 The Women's Elegance was Reflected in the 1930s

本章小结 Conclusions

1. 1920年代，受战争的影响，服装风格发生了翻天覆地的变化。
   Affected by the World War I, fashion styles in 1920s changed rapidly.
2. 1930年代的经济危机使服装的风格再次发生变化。
   Economic crisis in 1930s affected fashion styles again.

思考题 Thinking questions

1. 1920年代出现了几种服装风格，它们之间的区别是什么？
   There were several fashion styles in the 1920s. What are the differences between them?
2. 1920年代和1930年代服装风格在长度上有什么区别？
   What were the differences in length of garments between the fashion styles in the 1920s and 1930s?

基础理论

课题名称：20世纪中期的服装风格
Project name: Fashion Styles in the Mid-20th Century
课题内容：20世纪40年代服装风格，20世纪50年代服装风格，20世纪60年代服装风格
Course content: Fashion Styles in 1940s, Fashion Styles in 1950s, Fashion Styles in 1960s
课题时间：10学时
Project time: 10 hours
训练目的：详尽阐述服装从20世纪40~60年代的风格变化。
The purpose of training: Explain the changes in styles from the 1940s to 1960s.
教学要求：1. 使学生了解1940年代服装风格发生变化的原因。
    2. 使学生了解战争给服装带来的影响。
    3. 使学生了解1960年代超短裙流行的原因。
Teaching requirements:
  1. Make students understand the reasons why fashion styles changed so much.
  2. Make students understand the impact of war on fashion.
  3. Make students understand the reason why mini skirt was popular in 1960s.
课前准备：阅读相关服装美学和服装史方面的书籍。
Pre-class preparation: Read books on the history and fashion aesthetics of fashion.
课后练习：将20世纪前60年的不同服装风格整理出来，然后分析它们之间的不同点。
After-school exercises: Sort out the different styles of fashion in 1960s and analyze the differences between them.

第三章　20世纪中期的服装风格
Fashion Styles in Mid-20th Century

战争和时尚看起来截然不同。一个毁灭世界，一个创造了美丽。然而时尚，作为彻底的浮华，为法国人们的反抗提供了一个自然的出路。不管物质多么匮乏，规章制度多么严格，即使在第二次世界大战中，法国女人仍然当之无愧领衔世界时装潮流。其他地方的女人以穿着低调和朴素为义务，法国女人，却不顾一切地通过奢侈的穿着，表达了她们的独立性。

War and fashion looked very different, one destroy the world, and one create beautiful things. And yet fashion, the utter frivolity, provided a natural outlet for the defiance of the French citizens. No matter how scarce material was or how strict the rules and regulations, even during the Second World War French women lived up to their reputation as the best dressed in the world. Against all the odds, they expressed their independence by cultivating an extremely extravagant look. Even everywhere else women saw it as their duty to dress as inconspicuously and modestly as possible.

第一节　20世纪40年代服装风格
Fashion Styles In 1940s

1941年，德国所有的时装商店都被纳入了柏林，仅为出口而生产。在英国，从1941年开始，配给产生了最严格的规定：每件服装采购的材料，最大的裙长和裙宽，褶皱的最大数量，纽扣，饰品——一切都被严格控制。平民百姓被绝对禁止穿丝绸，因为它是用来做降落伞的。

由于美国政府把减少15%的

In 1941 all the German fashion houses were incorporated into Berlin, with production solely for export. In Britain, rationing from 1941 led to painfully exacting regulations: purchase of material per item of fashion, maximum skirt length and width, maximum numbers of pleats, buttons, and accessories—everything was regulated. Wearing silk was considered absolutely taboo among the civilian population, as it was used to make parachutes.

When the US government made it a national goal to reduce production of fabric by 15%, there was a "freeze" in

面料产量作为国家目标,时尚界被"冻结"了。他们猜测,如果风格不变,那么现有的时尚就可以流传下去。事实上,在1941~1945年时尚虽没有显著变化,但美国人在运动装方面取得了进展。具有讽刺意味的是,正是在战争期间,很多人才第一次感知到了质量。他们学会了欣赏耐用的材料,知道了好面料的皮肤触感,如棉花、羊毛和亚麻布。加工方法也突然变得很重要。一旦妇女们开始自己生产服装,她们马上能够分辨出工艺的好坏。

虽然美国人最早开了广泛穿着运动服装的先河,在着装上最崇尚自由,但在战后却也逐渐追求毫不考虑实用和价廉的高级时装。美国最重要的时尚杂志《哈泼斯》在战后就已经预期到美国时装市场和时装潮流的成熟和发展,这份杂志撰文说:"我们期待的不是丑小鸭,而是天堂鸟。"1947年的2月12日,克里斯汀·迪奥推出了他的第一个时装系列。虽然当时巴黎温度低于10摄氏度,但在蒙塔尼路的时装沙龙中,气氛却非常炽热。迪奥推出的设计无比典雅,窄小的腰部,优雅的曲线,宽大的裙子,震动了全场的来宾,他设计的上小下大的"A"字型裙,完全征服了整个巴黎的时装界。人们已经太久没有见过这种细腰丰臀造型的裙子了,在经历过两次世界大战之后,这种极具女性化风格的裙子再次出现在公众的视野中,并迅速流行起来。美国杂志《哈泼斯》的总编辑说:"这简直是一场革命。"迪奥的这场被称为"新风貌"的时装一出现就被

fashion. They conjectured that if style did not change, then available fashion could be handed down. In fact, between 1941 and 1945 there was no significant change in fashion, but the Americans moved ahead with sportswear. Ironically, it was during the war that many people first developed a feeling from quality. They learned to appreciate durable material that also felt good against the skin, such as cotton, wool, and linen. The way in which it was processed also suddenly became important. Once women began to produce everything themselves, they were able to tell good craftsmanship from bad.

Although America had been effectively free of fashion during the war, and had learned to live in casual, everyday clothes, it succumbed to the fascination of haute couture. This fulfilled the prediction made by the exclusive fashion magazine *HARPER'S* Bazaar at the end of the war, looking optimistically to the future: "We are waiting for birds of paradise, not little brown hens." On February 12, 1947, Christian Dior showed his very first collection and although the temperature outside were down to less than 10 degree, people were getting hot under the collar inside the salon on the Avenue Montaigne. Dior's design was incomparably elegant, with a narrow waist, elegant curves, and a large skirt that shocked the audience. His design of the small A-shaped skirt completely conquered the entire Paris fashion industry. It had been a long time since people hadn't seen this skirt with a tight waist and a hip shape. After two world wars, this feminine style of dress once again appeared in the public eyes and quickly became popular. Editor-in-chief of the American magazine *HARPER'S*, said: "This is a revolution." Dior's

女性们所追捧（图3-1）。

fashion, known as the "New Style", was sought after by women (Figure 3-1).

第二节　20世纪50年代服装风格
Fashion Styles in 1950s

20世纪50年代是高级时装的最后一个10年，也是典雅风格的最后一个时期。前无古人后无来者般的独立女服设计师出现，他们新奇的想法和高端的设计对时尚界产生了巨大的影响，时尚界彻底改变了方向。经过战时多年的简朴穿着和物资短缺，女人们渴望柔软的线条和奢华的大裙子。这个勇敢的新世界的确是美丽的！功能决定形式的现代主义的朴素线条，让位给了时尚且足具诱惑力的设计。沙漏造型，作为新风貌的标志外形，体现在生活的方方面面，大到建筑，小至内衣设计，甚至家里面最微不足道的物品，都能看见沙漏造型。火腿纹型的桌子、桶形座椅、郁金香杯、双头灯、沙漏形花瓶和曲线的玻璃烟灰缸，所有的设计都反映出"新风貌"那种优美的线条和形式。

新风貌带来了一系列典雅风格，紧随其后的一些款式特点是柔软的肩膀、圆润的臀部和蜂腰。直到20世纪50年代末，作为1920年代风格的回忆的几何形状开始重现，裙长到膝盖

The 1950s was the last great decade of haute couture. It was also the last period of elegant style. Independent women's wear designers who have never had before, their own novel ideas and high-quality designs had a huge impact on the fashion world. The fashion world was ripe for a radical change of direction. After years of clothes being austere and in short supply during wartime, women now yearned for soft lines and extravagantly full skirts. And this brave new world was indeed beautiful! The austere lines of modernism, in which form was dictated by function, gave way to attractive designs in a fashion geared towards seduction. The hourglass silhouette, which became a hallmark of the New Look, was echoed in every aspect of life, ranging from architecture, to interior design, to the most insignificant household object. Kidney-shaped tables, bucket seats, tulip glasses, two-headed lamps, hourglass-shaped vases, and curved glass ashtrays all mirrored the elegant line of the "New Look".

The New Look and subsequent variations were characterized by soft shoulders, rounded hips, and wasp waists. It was not until the end of the decade that geometric shapes, reminiscent of the 1920s, began to reappear and hemlines crept upwards to just under the knee. The shoes of this decade were narrow, pointed, and

第三章　20世纪中期的服装风格
Fashion Styles in Mid-20th Century

1940年代早期出现的军装元素
Military uniform elements appeared in the early 1940s

图 3-1

画出 20 世纪服装风格 Drawing the Fashion Styles in 20th Century

克里斯汀·迪奥的被称为
"新风貌"的服装
"New Look" garment designed by Christian Dior

图 3-1　20 世纪 40 年代服装风格
Figure 3-1　Fashion Styles in 1940s

以下。这时期的鞋是窄的、尖的，并且支撑在中高或理想的非常高的鞋跟上，它们逐渐变窄、变窄，直到变成了著名的"铅笔高跟鞋"或"细高跟鞋"。这一时期的帽子一般都较小，即使有着宽大的帽檐，帽顶也很扁平（图3-2）。

supported on medium-high or, ideally, very high heels, which gradually became narrower and narrower until they mutated into the famous "Pencil Heels", or "Stiletto". Hats of this period were generally small with a flat crow, even if the sported wide brims (Figure 3-2).

图 3-2

画出 20 世纪服装风格 Drawing the Fashion Styles in 20th Century

第三章　20世纪中期的服装风格
Fashion Styles in Mid-20th Century

图 3-2

画出 20 世纪服装风格 Drawing the Fashion Styles in 20th Century

图 3-2　20 世纪 50 年代服装风格
Figure 3-2 Fashion Styles in 1950s

第三节　20世纪60年代服装风格
Fashion Styles in 1960s

"摇摆的六十年代"无疑是20世纪最丰富多彩的十年，事实证明，直到今天，仍然没有关于这一时期的时尚特点的大致总结。有些人认为它是新自由的黄金时代，而另一些人则把它视为一个黑暗的十年，因为它见证了道德、权威和纪律的崩溃。然而，有一件事我们可以肯定的是，在这个时期播种的变革的种子至今仍然在社会、政治和文化的各个领域结成硕果。

一般来说，变革的动力来自年轻一代，来自最乐观的社会群体。由于战后的婴儿潮，年轻人占了总人口非常大的比例，他们的影响力也比以往任何时候都大。1950年代第一次被"发现"和认为是消费者的青少年，已长成二十来岁的叛逆者，质疑家长禀为神圣的所有一切。德国著名的模特乌施·欧泊梅尔回忆道："一切都是全新的：时尚、音乐、哲学——不用说还有我们的生活方式。我们不想和父母一代的生活有什么关系，我们的哲学是享受生活，尝试一切。"

一、小女孩风格

没有人确切知道究竟是玛丽·匡

The "Swinging Sixties" were undoubtedly the most colorful decade in 20th century, a view borne out by the fact that to this day there is still no general agreement about the merits of this period. Some people consider it the Golden Age of new freedom, while others regard it as a dark decade that witnessed the breakdown of morals, authority, and discipline. One thing we can be sure of, however, is that the seeds of change sown during that period are still coming to fruition to this day in various areas of society, politics, and culture.

The impetus for change stemmed from the younger generation—traditionally speaking, the most optimistic social group. Thanks to the postwar baby boom, young people accounted for an extremely large section of the population and their influence was greater than ever before. Teenagers, who had first been "Discovered" and courted as consumers in 1950s, had grown into rebels in their twenties, questioning all that parents held sacred. As Uschi Obermaier, a famous German model, recalled:"Everything was brand new: fashion, music, philosophy and—it went without saying—our way of life. We did not want sort of relationships our parents had. Our philosophy was to enjoy life and we experimented with everything."

Little Girlish Style

It is likely that no one will ever know for sure

特还是安德鲁·库雷热发明了超短裙，它就像突然在伦敦丢了一个炸弹。新时尚的到来引起了轰动的影响，把它看成新鲜和天真的学生裙，才能充分欣赏它的美。穿着超短裙的这些年轻女孩，都长着大眼睛和细长腿，傻傻地展示着她们刚发育的乳房和崭新的小女孩外貌。

在20世纪60年代，大多数女孩都想展示他们的小女孩气质，不仅是年轻人，连美国第一夫人杰基·肯尼迪也展示过她穿迷你裙的年轻气质，这使得《纽约时报》评论说"迷你裙的未来不言而喻。"杰基把精致的小西服、帽子和手套换成T恤衫和迷你裙，或产生牛仔裤，随着年轻人在伦敦的影响力越来越大，她对日趋年轻化的时尚潮流产生了巨大的影响。库雷热敢于大胆地把迷你裙引入高级时装界，把这件单薄的衣服改造成一种复杂的时尚。三角形的迷你裙给女人带来了活动的自由。库雷热还设计了和他的迷你裙匹配的短裤和长筒袜，这一切成为一个女人衣柜的基本元素（图3-3）。

二、波普风格

波普风格女装主要体现在色彩和图案上，用对比强烈富有冲击力的图形产生夸张奇特的表现效果。波普风格又称流行风格，它代表着20世纪60年代工业设计追求形式上的异化及娱乐化的表现主义倾向。从设计上来说，波普风格并不是一种单纯的一致

whether it was Mary Quant or Andre Cortèges who actually invented the miniskirt, which like a ground-breaking bomb originated in London. The sensational impact caused by the advent of the new fashion, looking as fresh and innocent as a child's pinafore dress, can only be fully appreciated. Along came these young girls, all huge eyes and long, thin legs, innocently revealing their budding breasts with the new little girlish look.

In 1960s, most girls wanted to show their little girlish look, not only the young person, but Jackie Kennedy, the first lady in US, showed her younger in miniskirts, which led *the New York Times* to comment that "The future of the miniskirts is now assured." Jackie's switching from tasteful little suits, hat, and gloves to T-shirts and a miniskirt, or jeans, inevitable contributed just as much to fashion's new, younger look as the growing influence of the young cult in London. It was a credit to Courrèges courage that he was bold enough to introduce the miniskirt into haute couture, transforming this flimsy garment into a sophisticated piece of fashion. The triangular-shaped mini dress gave women the freedom of movement. Courrèges also designed matching shorts and body stockings to wear with his miniskirts as basic elements of a women's wardrobe (Figure 3-3).

Pop Style

Pop Style dress is mainly reflected in the color and pattern, with a contrasting rich wallop graphics produces exaggerated peculiar performance effect. Pop Style, also known as the popular style, it represents the tendency of expressionism in form of alienation and entertainment in the sixties of the 20th century industrial design to pursue. From the design, Pop Style is not

第三章　20世纪中期的服装风格
Fashion Styles in Mid-20th Century

图3-3

069

第三章 20世纪中期的服装风格
Fashion Styles in Mid-20th Century

图3-3

071

第三章 20世纪中期的服装风格
Fashion Styles in Mid-20th Century

图3-3

073

画出 20 世纪服装风格　Drawing the Fashion Styles in 20th Century

第三章　20世纪中期的服装风格
Fashion Styles in Mid-20th Century

图 3-3

图 3-3 小女孩风格
Figure 3-3 Little Girlish Style

性的风格，而是多种风格的混杂。它追求大众化的、通俗的趣味，反对现代主义自命不凡的清高。在服装设计中强调新奇与奇特，并大胆采用艳俗的色彩，给人眼前一亮耳目一新的感觉刺激。"波普"是一场广泛的艺术运动，反映了战后成长起来的青年一代的社会与文化价值观，力图表现自我，追求标新立异的心理。波普风格在服装中的表现是：追求大众化、通俗化的趣味，追求新颖、追求古怪、追求稀奇，设计中强调新奇与独特，采用强烈的色彩处理（图3-4）。

三、太空风格

人类征服外太空是20世纪60年代的重大新闻。当时美国宇航员尼尔·奥尔登·阿姆斯特朗第一次踏上月球，这似乎真的标志着一个新时代的到来，同时带来了服装上的变化。

第一位推出未来主义时装系列的设计师是安德鲁·库雷热，他带来了一系列太空风格的设计。他的设计涉及长腿"月亮女孩"。1964年，她穿着库雷热的白色长裤和短裙，穿着低跟羊皮靴，走上了T台。库雷热喜欢强力织物，如马裤呢、华达呢和双面羊毛，这种面料不会跟随身体运动。他采用建筑设计手段，裁剪出一系列具有建筑感外观的作品。"我们总是在两层织物之间插入衬里来产生体积，"库雷热解释道。每件设计的起点都是肩膀，因为那里没有折叠、皱

simply a consistent style, but a variety of mixed style. The popular taste it, the pursuit of popular opposition, modernism pretentious. In the fashion design it emphasizes the novelty and strange, and bold use of color, the color of the light, giving people a bright and refreshing feeling of stimulation. "Pop" is a broad movement of art, reflecting the grew up in the post-war young generation of social and cultural values, and tries to express themselves and to pursue the unconventional and psychological. Pop style in fashion design : pursuing popularization, popularization of the interest ,pursuing of novelty and eccentricity, the design emphasizes new and unique, with strong color processing (Figure 3-4).

Space Style

Man's conquering of outer space was big news in 1960s. when American astronaut Neil Alden Armstrong set foot on moon for the first time, that really seemed to mark the dawn of a new age, bringing with it changes in our clothes.

The first designer to launch a futurist fashion line was Andre Courrèges with his space look. Courrèges' vision involved long-legged "Moon Girls", who he sent out onto the runway in 1964 wearing white trousers and short skirts, worn with low-heeled goatskin boots. Courrèges favored strong fabrics such as whipcord, gabardine, and double-faced wool, which did not follow the body when one move. He used architectural design to cut out a series of architecturally appealing works. "We always inserted lining between two layers of fabric in order to create volume," explained Courrèges. The starting point of every design was the shoulders since there were no tucks, gathers, or even draped effects,

第三章　20世纪中期的服装风格
Fashion Styles in Mid-20th Century

图 3-4

第三章 20世纪中期的服装风格
Fashion Styles in Mid-20th Century

图 3-4 波普风格
Figure 3-4 Pop Style

褶甚至垂褶效果，只有平行线、水平线或垂直线。他利用弯曲或圆形的贴袋和厚厚的绲边来柔化衣服的朴素的几何形状。唯一的装饰是用来突出服装结构的可见的明线缝合线（图3-5）。

四、嬉皮士风格

在20世纪60年代，音乐是超越国家、阶级和性别界限并联合西方世界年轻人的元素。在音乐方面，比尔·海利和埃尔维斯·普雷斯利是20世纪60年代音乐的先驱，后来由甲壳虫乐队主宰。时尚的模特留着他们的"披头士"发型，很快就被标记为嬉皮士风格。嬉皮士对塑料的厌恶和对使用天然材料的喜爱广泛流行起来。

追求无拘无束，自由自在的生活方式是嬉皮精神的实质。嬉皮士风格女装呈现自由、随意的效果，在图案、色彩、材质、装饰手法等方面喜欢将各地区、各时代的民族风格服装组合在一起，形成怀旧、浪漫和自由的设计风格，并带有相当的异域情调。嬉皮士追求自由的生活方式，东方的宽松服装深受他们的喜爱，同时混合各民族服饰元素是嬉皮士风格女装的主要特点，如手工缝制、手工印染等。具有阿拉伯和东方情调的流苏是嬉皮士风格服装的一个主要装饰手法。圣罗兰是一个典型的嬉皮士风格的时装设计师。他的晚礼服把复古风格和民族风格混成嬉皮士风格。他声称，"在我心里，晚上是民俗时间。"他的设

merely parallel lines, either horizontal or vertical. The austere geometry of his clothes was softened by curved or rounded patch pockets and thick piping. The only decoration consisted of visible topstitched seams, which highlighted the garment's construction (Figure 3-5).

Hippies Style

In 1960s, music was the element that transcended national, class, and gender boundaries and united the young people of the Western world. In music terms, Bill Haley and Elvis Presley were precursors of the Sixties' music, and later was dominated by the Beatles. The stylish Models with their "Beatle" haircuts, very quickly were labeled as Hippies. Hippies' aversion to plastic and commitment to the use of natural materials were widely adopted.

The pursuit of unfettered and free way of life is the essence of the hippy spirit, also the Hippies Style women's clothing presents a free and random effect. In the aspects of patterns, colors, materials, decorative techniques, etc, they like to combine ethnic styles of various regions and eras to form a nostalgic, romantic and free design style, quite exotic. Hippies in pursuit of free way of life, the east of loose fitting clothes by their love, mostly mixed ethnic costume elements is the main characteristics of women Hippies Style, such as hand sewing, hand dyeing. Arabia and Oriental fringe are main decoration of Hippies style fashion. Saint Laurent was a typical fashion designer of Hippies Style. His evening wear reveled in the retro and ethnic looks so beloved of the Hippies Style. "In my mind," he admitted, "evening is

第三章　20世纪中期的服装风格
Fashion Styles in Mid-20th Century

图3-5

画出 20 世纪服装风格 Drawing the Fashion Styles in 20th Century

第三章　20世纪中期的服装风格
Fashion Styles in Mid-20th Century

图 3-5

画出 20 世纪服装风格 Drawing the Fashion Styles in 20th Century

第三章 20世纪中期的服装风格
Fashion Styles in Mid-20th Century

图3-5

087

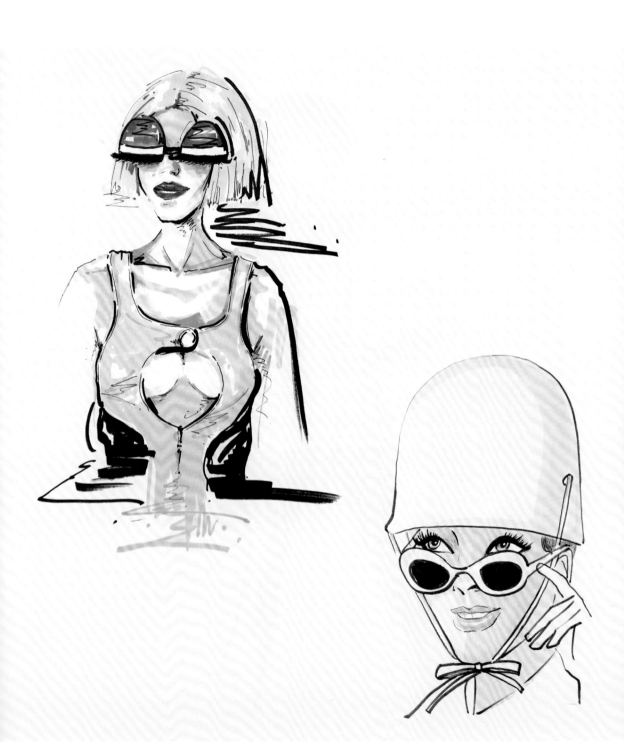

图 3-5 太空风格
Figure 3-5 Space Style

计灵感来自古老的中国、秘鲁、摩洛哥、中部非洲或优雅的威尼斯卡萨诺瓦时代（图3-6）。

the time for folklore." His designs were inspired by ancient China, Peru, Morocco, Central Africa, or the elegant Venice of Casanova's times (Figure 3-6).

图3-6

第三章　20世纪中期的服装风格
Fashion Styles in Mid-20th Century

图 3-6　嬉皮士风格
Figure 3-6 Hippies Style

本章小结 Conclusions

1. 1950年代之后，服装风格变得多样化。

    More and more fashion styles appeared after 1950s.

2. 1960年代的服装摆脱了各种禁忌。

    Fashion in 1960s got rid of many taboos.

思考题 Thinking questions

1960年代的口号是年轻化，导致这种年轻文化流行的原因是什么？

What made youth as the slogan of 1960s?

基础理论

课题名称：20世纪后期的服装风格
Project name: Fashion Styles in the Late 20th Century
课题内容：20世纪70年代服装风格，20世纪80~90年代服装风格
Course content: Fashion Styles in 1970s, Fashion Styles From the 1980s to the 1990s
课题时间：10学时
Project time: 10 hours
训练目的：详尽阐述服装从20世纪70~90年代的风格变化。
The purpose of training: Explain the changes in style from the 1970s to 1990s.
教学要求：1. 使学生了解1970年代服装风格多样性的原因。
　　　　　2. 使学生了解1980年代服装风格的特点。
　　　　　3. 使学生了解解构风格的主要特征。
Teaching requirements:
　　　　1. Make students understand the reasons there were kinds of styles in 1970s.
　　　　2. Make students understand the characteristics of fashion styles in 1980s.
　　　　3. Make students understand the characteristics of Deconstruction Style.
课前准备：阅读相关服装美学和服装史方面的书籍。
Pre-class preparation: Read books on the history and fashion aesthetics of fashion.
课后练习：将20世纪70~80年代的不同服装风格整理出来，然后分析它们之间的不同点。
After-school exercises: Sort out the different styles of fashion in 1970s and 1980s, then analyze the difference between them.

第四章　20世纪后期的服装风格
Fashion Styles in the Late 20th Century

第一节　20世纪70年代服装风格
Fashion Styles in 1970s

　　头发插花，脚上穿着耶稣式凉鞋，嘴唇挂着微笑：1960年代的年轻理想主义者飘然进入1970年代。他们的乌托邦似乎成了一个现实，未来属于年轻人和"爱与和平"的圣歌。在现实中，虽然这种年轻人狂热的宗教崇拜保留了，引发了这种崇拜的年轻人却老了。自然地，他们的成就和启蒙的源泉，现在转向反对他们，因为他们的格言是"不要信任超过30岁的人"，留着胡子和长发的温和嬉皮士很快就看起来过时了。年轻的追随者在剩余的时间内也没有得到祝福，痛苦地经受了失业、通货膨胀和无聊；因为年轻并不自然意味着理想主义和乐观。

　　政治团体并不是唯一试图让他们的理想成为现实的团体。曾经被压迫的女性，为了实现个人成就，不惜牺牲整个家庭而去工作，这导致了很大的不确定性。于是突然间，一切开始围绕着个人的欲望。人们关心自己，漠视公共利益，记者汤姆·沃尔夫把20世纪70年代称之为的"自我的十年"。

　　Flowers in their hair, Jesus sandals on their feet, and a smile on their lips: the young idealists of the 1960s wafted happily into the seventh decade of the 20th century. Their utopia seemed to become a reality. The future belonged to the young and their "Love & Peace" mantra. In reality, while this cult of youth lived on, the young people who had sparked the change just got older. Naturally, their source of fulfillment and enlightenment, now turned against them. The placid hippies with their beards and long hair soon began to look passe, and before long the watchword was "Trust no one over 30." The young people who came after them did not see their lesser years as a blessing, suffering as they did from unemployment, inflation, and boredom; being young did not automatically mean being idealistic and optimistic.

　　Political groups were not the only ones trying to make their ides a reality. People who had been oppressed, predominantly women, now strove for personal fulfillment, and whole families often fell by the wayside. This led to great uncertainty, as all of a sudden everything revolved around the desires of the individual. Everyone cared about himself and disregarded for the common right, the journalist Tom Wolfe aptly

"我们想造反，"薇薇安·韦斯特伍德解释说："我们觉得嬉皮士运动应该顺其自然，我们从来就不喜欢把它摆在首位。"她和她的合作伙伴马尔科姆·迈凯轮经营了一个伦敦高档摇滚风格的女装店，那里充斥着摇滚和紧身裤，是20世纪50年代地下青年文化的发起地。后来改卖皮革、乳胶、束身衣，一直到韦斯特伍德最终被称为"朋克之母"。

作为一个品味较差的时代，这十年已经过去了，平台高跟鞋和热裤，喇叭裤和涤纶衬衫，迪斯科闪光面料，复古媚俗以及未来的朋克；一切都被抽样、混合、拒绝，并再次被接受。这可能被视为仅仅是抗议的表达，但它掩盖了一种解放的，创造性的本能，其影响至今仍可以看到。从1973年开始，时尚就像生活一般，不再受固定规则的约束；每个人都可以穿着适合自己的衣服。正如我们今天所做的那样，将各式不同的衣服混搭在一起，也是嬉皮士的遗风，他们通过自由穿着来表达自己的个性。

一、迪斯科风格

白天穿着不起眼的灰色和米色受人尊敬的职场人士，到了晚上着装也开始变得浮华和夸张。厌倦了古板的商务套装，迪斯科服装提供了一剂良药。引人注目的涤纶衬衫，莱卡紧身衣裤，露背上衣搭配金银纱热裤，蕾丝衬衫配上闪闪发亮的牛仔裤，1940

dubbed the 1970s the "Egotistical Decade".

"We want to rebel," explained Vivienne Westwood, "As we felt that the hippie movement had run its course, and we had never liked what it had brought with it in the first place." Together with her partner Malcolm McLaren, she ran the London boutique "Let it Rock", where the underground youth culture of the 1950s lived on, with rock 'n' roll and drainpipe trousers. It later switched to selling leather, latex, and bondage fashion, until Westwood eventually became known as "Mother of Punk".

This decade has gone down as the era of bad taste for a reason. Platform heels and hot pants, flares and polyester shirts, disco glitter, retro kitsch, and future punk; everything was sampled, mixed, rejected, and taken up again. This might be dismissed as a mere expression of protest, but it concealed a liberating, creative instinct whose effects can still be seen today. From 1973, fashion, like life in general, was no longer subject to fixed rules; everyone could dress in what suited them. Putting together a wardrobe of separates, as we do today, is also a legacy of hippies, who expressed their individuality by dressing as they pleased.

Disco Style

In the evening even respectable professionals who dressed in unassuming gray and beige by day embraced glitz and excess. Bored of their prim business outfits, the disco provided the antidote for good taste. Shrieking polyester shirts, lycra body stockings, halter neck tops paired with silver lurex hot pants, lace shirts with glitter jeans, artificial silk dresses from 1940s,

代的人造丝质连衣裙，1950年代的鸡尾酒会礼服，祖母式的拖地华丽长袍，现代的晚礼服，开口直至胯部；甚至是赤裸裸的身体，一切经过精心的装扮后都可能出现。迪斯科，曾经是同性恋者的专属，成为展示自我的舞台，按照安迪·沃霍尔的信条，每个人都可以成名十五分钟。数量庞大的颜色和材料产生了冲击——"最好的品位是坏品位"是那时候的座右铭（图4-1）。

二、朋克风格

朋克风格是20世纪70年代最激进的时尚风格。他们剃胡子，文身，穿一身被认为难看的有孔的衣服：垃圾袋、撕破的T恤、黑色皮革、闪闪发光的面料、豹皮、军装……他们刻意表现出的无品味，有点像老迪斯科一代。30多岁的年轻人平静地把难看的、可笑的时尚视为有趣和迷人的，朋克们的目标则是痛苦的挑衅。他们怒气冲冲地向前看，发现"没有未来"。朋克们用"性与暴力"取代了"爱与和平"，用令人不安的人造物取代了自然崇拜。服装由决不属于同类的元素组成：稚气的面孔、马桶链、卫生棉球和作为皮夹克装饰品的安全别针、纳粹党徽和头骨，以及从性用品商店买来的荧光蕾丝内衣。伴随着这些令人惊讶的不协调东西的混搭，他们这种疯狂的组合常常呈现出一种超现实主义的特征，让人联想到达达主义运动，更重要的是，当像维维安·韦斯特伍德这

cocktail dresses from the 1950s, grandma's tweed floor-length gowns, or modern evening dresses, slit to the crotch; everything could be put on for show, even the naked body—artfully painted, of course. Disco, once the sole preserve of gays, became the stage for all self-exhibitionists and following of Andy Warhol's creed, that everyone could be famous for fifteen minutes. Crazy population of colors and materials formed a hit—"The best taste is bad taste" was the motto of that time (Figure 4-1).

Punk Style

Punk Style was the most radical fashion style in 1970s. They clad their shaved, tattooed, pierced bodied in everything that was considered ugly: garbage bags, ripped T-shirts, black leathers, glittery fabric, leopard-skin, military fashion…In their deliberate tastelessness they bore some resemblance to the older disco generation, but while the over 30s serenely saw unsightly, ridiculous fashion as amusing and charming, the punks aimed at bitter provocation. They looked ahead in anger, discovered that there was "No Future". The punks replaced "Love & Peace" with "Sex & Violence", and nature worship with jarring artificiality. Outfits were composed only of elements that did not belong together: childish faces; toilet chains, tampons, and safety pins as adornment for leather jackets, along with swastikas and skulls, over fluorescent lace underwear from sex shops. With this surprising juxtaposition of incongruous items, their crazy combinations often took on a surreal character, reminiscent of the Dada movement, and the most important was, when someone as shrewd as Vivienne Westwood was behind it, they

第四章 20世纪后期的服装风格
Fashion Styles in the Late 20th Century

图4-1

画出 20 世纪服装风格 Drawing the Fashion Styles in 20th Century

图 4-1　迪斯科风格
Figure 4-1 Disco Style

样精明的人在背后时，他们对如何使朋克风格流行有着深刻的理解（图4-2）。

三、中性风格

为了表现中性风格，牛仔裤的选择变得很微妙，蓝色粗斜纹牛仔裤到处都是磨损，无论男人还是女人、男同性恋者和女同性恋者、富人和穷人都是如此。男孩和女孩看起来同样悲惨。他们穿着上大学的课程或去办公的牛仔裤——不再是1960年代大获青睐的有花边刺绣的、喇叭形的牛仔裤，而是褪色的、像被珍爱多年的牛仔裤。许多人穿着看起来好像从来没有换过的衣服，因为这是证明他们不关心时尚的最好的方式。然而，这波双性同体浪潮仅仅是平等的行为，并没有得到所期望的平等权利（图4-3）。

四、波西米亚风格

波西米亚风格是指一种保留着游牧民族特征的服装风格，其特点是鲜艳的手工装饰和粗犷厚重的面料。层叠蕾丝、蜡染印花、皮质流苏、手工细绳结、刺绣和珠串，都是波西米亚风格的经典元素。波西米亚风格代表着一种前所未有的浪漫化、民俗化、自由化。也代表一种艺术家气质，一种时尚潮流，一种反传统的生活模式。波西米亚服装提倡自由和放荡不羁及叛逆精神，浓烈的色彩让波西米亚风格的服装给人以强烈的视觉冲击感。波西米亚为Bohemian

had a brilliant understanding of how to make punk styles fashionable (Figure 4-2).

Neutral Style

To show the Neutral Style, the choice of jeans became tricky, and the blue denim pants were worn everywhere, by men and women, gays and lesbians, rich and poor. Boys and girls looked equally wretched and miserable. They could be worn to university classes or the office—not the embroidered, flared version favored in 1960s, of course, but a faded pair that looked as though they had been much loved by years. Many of those wearing them looked as though they never changed their clothes, as this was the best way of demonstrating they cared little about fashion. Yet this wave of androgyny led not to the hoped-for equality of rights, but to equally of behavior (Figure 4-3).

Bohemian Style

Bohemian Style refers to a style that preserves the characteristics of nomadic people, characterized by bright hand-decorated and rough and heavy fabrics. Laminated lace, batik prints, leather tassels, hand-knot knots, embroidery and beads are all classic elements of Bohemian Style. The Bohemian Style represented an unprecedented romantic, folklore and liberalization. It also represented an artist temperament, a fashion trend, an anti-traditional lifestyle. Bohemian clothing promotes freedom and debauchery and rebellious spirit, and the strong colors give Bohemian Style clothing a strong visual impact.

图 4-2 朋克风格
Figure 4-2 Punk Style

第四章 20世纪后期的服装风格
Fashion Styles in the Late 20th Century

图 4-3 中性风格
Figure 4-3 Neutral Style

的译音，原意指豪放的吉卜赛人和颓废派的文化人。追求自由的波西米亚人，在浪迹天涯的旅途中形成了自己的生活哲学。波西米亚不仅象征着拥有流苏、褶皱、大摆裙的流行服饰，更是自由洒脱、热情奔放的代名词。波西米亚风格的服饰是一场革命。

波西米亚风格的特点是：

不羁。多褶大摆裙、蜡染印花、皮质流苏、手工细绳结、刺绣和珠串，神秘华丽的波西米亚风格将你装扮成丛林精灵，如敏感灵动的小鹿，摸不着、抓不住、猜不透，野得潇洒自在。现实生活中，如麋鹿般，印花、蜡染的宽松裙装，烦琐层叠的首饰搭配，浪漫华美的波西米亚风情让许多明星少了冷漠距离感，多了份优雅与亲切。

格调。波西米亚的装扮，逃不了一条打满粗褶细褶的长裙，它可以是纯棉的、粗麻的、砂洗重磅真丝的，可以是镂空设计的、缀满波西米亚式绣花的、加上婀娜的荷叶边的、垂垂吊吊满是流苏的，可以是布满无规则图案的、用其他风格面料拼镶的……总之它是繁复的、奢华的，时刻都在昭示着自己的独特，它让穿上它的女性霎时间变得超凡脱俗并蔑视一切。如果还要披上外套，那最好是一件收腰收得恰到好处的长大衣，昂贵的羊绒当然是第一选择，退而求其次便是精纺亚麻，加一条粗犷而帅气的腰带，将硬朗与柔美完美地结合起来。

饰品。要做个地道的波西米亚女郎，你最好不要放过身体上任何能披

Bohemian is the transliteration of Bohemian, originally meant the uninhibited Gypsies and the decadent cultural people. The Bohemians who pursued freedom formed their own philosophy of life in the journey of the world. Bohemian not only symbolizes the fashions with tassels, folds, and large swing skirts, but is also synonymous with freedom and enthusiasm. Bohemian Style clothing is a revolution.

The Bohemian Style is characterized by:

Uninhibited. Multi-pleated skirt, batik print, leather tassels, hand-knot knots, embroidery and beads, the mysterious bohemian style will dress you up as a jungle elf, such as a sensitive deer, can't touch, can't catch, can't guess, wild and comfortable. In real life, elk, print, batik loose dresses, cumbersome cascading jewelry, romantic and bohemian style, many stars have less indifference, more elegant and friendly.

Style. The bohemian dress can't escape a long skirt with thick pleats and pleats. It could be cotton, numb, sand-washed heavy silk, or hollowed out, embellished with bohemian embroidered, with a lotus leaf, hanging hangs full of tassels, or covered with irregular patterns, mosaic with other styles of fabrics... In short, it is complex, luxurious. Every moment is a sign of its own uniqueness, which makes the woman who wears it a time to become extraordinary and despise everything. If you still want to put on a jacket, it is best to have a long coat that is just right. The expensive cashmere is of course the first choice. The second best choice is the worsted linen, plus a rough and handsome belt. The combination of toughness and femininity is perfect.

Accessories. To be an authentic bohemian girl, you had better not let go of any part of the body that

挂首饰的部位，手腕上、脚踝上、颈前、腰间，还有耳朵、指尖，别人戴一串，你戴三串，别人挂细的，你就挂粗的，这两年疯狂流行的藏饰被波西米亚女郎们引为至宝，那些发黑的银器、天然的或染色的石头，不管它重不重、贵不贵，统统往身上、手上套了再说。走动间，一定要浑身上下叮当作响；点烟时、端起一大扎啤酒时，一定要让连着戒指与手镯的链子斜斜垂下（图4-4）。

can be worn with jewelry, on the wrist, on the ankle, on the front of the neck, on the waist, on the ears, at the fingers, and on others. You would wear three strings if others hangs one; you are hanging thick if others hangs thin .Crazy popular Tibetan ornaments in the two years are cited as treasures by Bohemian girls, those blackened silverware, natural or dyed stones, even it is heavy, expensive, and it is all over the body. When you walk, be sure to squeak the body; when you smoke, when you are holding a big beer, you must let the chain with the ring and bracelet slant down (Figure 4-4).

图4-4

画出 20 世纪服装风格 Drawing the Fashion Styles in 20th Century

第四章　20世纪后期的服装风格
Fashion Styles in the Late 20th Century

图4-4

图 4-4　波西米亚风格
Figure 4-4　Bohemian Style

第二节　20世纪80~90年代服装风格
Fashion Styles From the 1980s to the 1990s

有关这个话题的畅销书出现在1979年，成了那些时尚尖端者的圣经。职业女性逐渐获得了社会地位。女员工像男同事一样使用新的现代的个人电脑技术——苹果电脑，手机和备忘记事本，成了她们最喜欢的饰品，强调作为一个女商人的重要性比任何女性的装饰品更有效。最重要的是，女性学会了如何将时尚作为职业成功的一种手段，女套装裤成了高层女性员工的标准工作服。然而，在对她们施加更严格的着装规定之前，女性几乎没有人闯入她们职业的上层。如果她们和男人共事，她们不得不放弃裤子改穿裙子。及膝的黑丝绸连衣裙配以古板的西装，晚上必须佩戴珠宝，女性活动的范围从办公室延伸到宴会，这是一个突破，这种做法本身就是革命性的。成功的女人通常就如《成功着装》杂志上建议的：穿西装裙和一件丝绸衬衫、肉色丝袜、低跟鞋和低调的黄金首饰。

20世纪80~90年代的伟大承诺是"人人皆奢侈"，这似乎听起来像高级定制时装结束的信号。曾经提供专用服务的时装屋越来越多地开始转变成出售一切物品的商店，从香水到太阳镜。曾经一度是社会精英

A bestseller with this title appeared in 1979 and became a bible for those on the up. Career women were gradually gaining ground. Female employees used the new personal computer technology pioneered by Apple just as much as their male colleagues, and the cell phone and Filofax, became their favorite accessories, emphasizing a businesswoman's importance more effectively than any feminine adornments. Above all, women learned how to deploy fashion as a means to professional success and the pantsuit became standard work wear for female employees in top positions. Yet women had scarcely broken into the upper echelon of their professions before an even stricter dress code was imposed on them. If they were to play along with the guys, they had to abandon their pants and go back to wearing skirts. Knee-length dark silk dresses that could be worn with the obligatory blazer and dressed up with jewelry in the evening, allowing women to go from the office to evening events, were a hit, and this approach itself was revolutionary. The successful women generally wore a skirt suit and a silk blouse, nude stockings, low-heeled shoes and subtle gold jewelry, as recommended by *Dress for Success*.

The great promise of the 1980s—1990s was "Luxury for all", and it seemed to sound the death knell for haute couture. Once-exclusive fashion houses increasingly began to transform themselves into stores that sold everything that make money, from scent to sunglasses. What had once been the exquisite privilege of the cream

的精致特权，如今却变成了大众消费。品牌变得比款式本身更重要，而选择正确的商标则代表了好的品位。身份地位标志被掩盖；质量被忽略，对于想挤入上层社会的人来说，"成功着装"是神奇的方法（图4-5）。

of society now became subject to mass consumerism. The brand became more important than style itself, and the hunt for the right logo supplanted good taste. Status symbols were covered; quality was irrelevant. The magic formula for social climbers was "Dress for Success" (Figure 4-5).

图 4-5
Figure 4-5

一、雅皮风格

社会的最新角色模型是"雅皮士"（城市年轻职业人士），他们是计算机技术或媒体职业从业者，或是做股票市场赚钱的喜欢炫耀性消费的金融业人士。20世纪80年代，《达拉斯》和《迈阿密的罪恶》这两个电视剧系列，深深影响了人们的生活。在电视系列里，女士们穿着考究像男装似的，卷着衣袖的皱巴巴的亚麻夹克，唐·约翰逊和菲利普·米迦勒·托马斯多次穿着这样由尼诺·切瑞蒂设计的折痕清晰休闲优雅的裤子。这位充满魅力的意大利人是20世纪80年代的好莱坞最受欢迎的设计师。雅皮士引导的奢侈文化终止于1987年10月股票市场崩溃后（图4-6）。

二、解构风格

解构主义最早出现在建筑中，其思想打破了传统造型与陈旧的观念，在20世纪80年代末兴盛起来并应用到服装领域，至今仍对服装设计影响广泛。解构主义设计在意识形态和设计理念上与传统思维模式相比别具一格，具有突破性和创新性，形成了一种前卫的设计理念。解构主义设计师对现代主义设计和后现代主义的设计形式皆表现出非常不满。避免对称、破损散乱、残缺、突变、失重和超长等特征是解构主义风格在服装设计中运用的主要表现形式。解构主义的形象特征

Yuppie Style

Society's latest role models were the "yuppies" (young urban professionals), who had careers in computer-based technology or the media, or else made lots of money playing the stock market and casually spent or all again on conspicuous consumption. In 1980s, two TV series, *Dallas* and *Miami Vice*, affected people's life deeply. In the series, the ladies dressed elegantly similar as male counterparts, the crumpled linen jackets with rolled up sleeves and sharply creased trousers worn with such casual elegance by Don Johnson and Philip Michael Thomas were created by Nino Cerruti. This charismatic Italian was Hollywood's most popular designer during the 1980s. The Yuppies and its luxury-oriented culture was sealed when the stock market crashed in October 1987 (Figure 4-6).

Deconstruction Style

Deconstruction first appeared in architecture with the idea of breaking the traditional shape and obsolescence, and it flourished in the clothing field in the late 1980s and still has a wide influence on fashion design. Deconstruction design was unique in its ideology and design concept compared with the traditional mode of thinking. It was groundbreaking and innovative, and formed an avant-garde design concept. Deconstructive designers were very dissatisfied with the design of modernism and postmodernism. Avoiding symmetry, damage, disintegration, incompleteness, mutation, weightlessness and super-long features are the main expressions of Deconstruction

画出 20 世纪服装风格 Drawing the Fashion Styles in 20th Century

第四章　20世纪后期的服装风格
Fashion Styles in the Late 20th Century

图 4-6

画出 20 世纪服装风格 Drawing the Fashion Styles in 20th Century

图 4-6 雅皮风格
Figure 4-6 Yuppie Style

主要有：打破常规，防止出现常见的完整的对称的结构，巧妙转移服装中某个部位的结构，使服装整体外观造型显得支离破碎、凌乱、矛盾（图4-7）。

三、极简主义风格

在1990年代中期，时尚界出现了一种被称为极简主义的基本款式。许多人对时尚失去了兴趣。这种美国零售商称之为万无一失的服装随处可见，就像那个时代一样，实用、微妙、端庄。这些基本要素包括经典裁剪的外套、裤装、铅笔裙以及20世纪50年代出现的服装类型。极简主义主要外观是：高领毛衣，最好是黑色的。基本风格的服装很快填满了商店的柜台，占据了即使是最简单的东西也有奢华触感的秀场。用最好的面料，裁剪更加复杂，几乎盲目地关注细节。事实上，这种风格的服装看起来设计简单，制作却复杂且价格昂贵，富有的顾客喜欢这种低调的奢华。正如卡尔·拉格菲尔德所说："奢侈就是看不见的东西和看得见的东西一样精美，真正的奢华就藏在里面。"

严格而感性的极简主义风格是通过时尚、优雅的黑色设计和后来的灰色设计实现的。除了纯白经典衬衫外，其他颜色和图案都被嗤之以鼻。人们重新发现了"好品位"，

Style used in fashion design. The image features of deconstruction mainly include: breaking the routine, preventing the appearance of common complete and symmetrical structures, and subtly transferring the structure of a certain part of the garment, making the overall appearance of the garment appear fragmented, messy, and contradictory (Figure 4-7).

Minimalism Style

In the mid-ninties, fashion turned into a basic style known as minimalism. Many people lost their interest in fashion. While basic, as American retailers dubbed any failsafe garment, were worn everywhere, as they were practical, subtle, and demure, like the era itself. These basics included blazers with classic tailoring, pantsuits, pencil skirts, and that staple of existentialists in the 1950s, celebrated for its minimalist look: turtlenecks sweaters, preferably in black. Basic soon filled the rails in the shops and dominated the runways, where even the simplest pieces had a luxurious touch. Only the best materials were used, the tailoring became even more sophisticated, and there was an almost fetishistic attention to detail. Designs that looked simple were, in fact, extremely complex and expensive to produce, but wealthy costumers allowed themselves a little hidden extravagance. Just as Karl Lagerfeld said: "Luxury is when what is invisible is as exquisite as what is visible. The real luxury is in the lining."

The strict yet sensual aesthetic of Minimalism Style was achieved with chic, elegant designs in black, and later gray. Other colors and patterns were frowned upon, with the exception of the pure white classic shirt. People rediscovered "Good Taste," and while the wealthy rejoiced in it, many others basked in the good conscience that came

画出 20 世纪服装风格 Drawing the Fashion Styles in 20th Century

第四章　20世纪后期的服装风格
Fashion Styles in the Late 20th Century

图 4-7　解构风格
Figure 4-7 Deconstruction Style

虽然富人对此感到高兴，但更多人却是依靠良知，来选择自然色的识时务的"生态"外观，遵循"少即是多"的原则。昂贵、极简的设计成了绝对排他性的代名词。着装越简洁越好；没有一点多余的东西，那就不会出错（图4-8）。

from opting for a politically correct "Eco" look in natural colors, following the "Less is More" mantra. Expensive, minimalist design became a byword for utter exclusivity. The more pared down the outfit, the better; if there is nothing superfluous, than nothing can go wrong (Figure 4-8).

第四章　20世纪后期的服装风格
Fashion Styles in the Late 20th Century

图4-8

画出 20 世纪服装风格　Drawing the Fashion Styles in 20th Century

图 4-8 极简主义风格
Figure 4-8 Minimalism Style

本章小结 Conclusions

1. 1970年代也是20世纪服装史中风格出现较多的时期。

    The 1970s was also a period of more fashion styles in the 20th century.

2. 1980年代的服装风格在廓形上呈倒三角形。

    The fashion styles in 1980s was an inverted triangle in the silhouette.

3. 1990年代的解构风格是对服装的打破和重组。

    The Deconstruction Style appeared in 1990s was the breaking and restructuring of clothing.

思考题 Thinking questions

1. 1980年代服装呈倒三角形原因是什么？

    What was the reason for the appearance of an inverted triangle in the 1980s?

2. 是什么原因导致解构风格的出现？

    What made Deconstruction Style appear?

应用实践

课题名称：服装风格的时装画表现及欣赏
Project name: Performance and Appreciation of Fashion Drawing
课题内容：通过对不同风格时装画的欣赏，讲解绘画表现技法和注意事项
课题时间：14学时
Project time: 14 hours
训练目的：欣赏时装画的画法，尤其是马克笔时装画的表现技法
The purpose of training: appreciating the fashion drawing, especially drawing by marker.
教学要求：1. 使学生掌握不同风格时装画的画法
　　　　　2. 使学生了解服装画层次的表达
Teaching requirements:
　　　　1. Make students grasp the ways to draw different fashion styles.
　　　　2. Make students understand the ways to draw the layers.
课前准备：阅读相关服装画的书籍。
Pre-class preparation: Read books on the fashion drawing.
课后练习：训练用马克笔画不同风格的服装
After-school exercises: Training to draw different fashion styles with markers.

第五章 服装风格的时装画表现及欣赏
Performance and Appreciation of Fashion Drawing

在马克笔时装绘画中，一般的流程是：先进行线稿大致定型——通过勾线再明确整体形状和细节部分——运用浅色马克笔大致铺出面部和头发底色——用过渡色对整幅画进行上色——用重色马克笔再次对画面进行上色——最后运用一些补色和技法对画面整体进行调整和完善。

注：线稿阶段需要清晰表现人物整体动态，画面的布局安排，先在脑子里想一下最后希望呈现出来的大致效果。勾线阶段需要对细节部分进行合理定位，运用些手法让画面整体显得流畅又保持形状。

浅色马克笔阶段需要大胆地对面部和服装进行铺色，可以利用对人体结构的理解进行铺色。过渡色阶段需要胆大心细画对转折位置，重色是对画面进行接地气的阶段，不让整幅画显得太飘，合理大胆地利用重色进行压色，体现层次的同时让画面具有空间感。调整阶段，也是让这幅画体现个人特质阶段，需要结合自身想法进行调整。背景部分同样需要合理规划，通过对主体色彩的呼应选取适当的颜色进行上色。

When we use markers, the general process is: firstly, the line draft is roughly shaped - through the hook line to clear the overall shape and details - using a light-colored marker to spread the face and hair background roughly-coloring the entire painting with a transition color – coloring the picture again with a heavy-colored marker – finally using some complementary colors and techniques to adjust and perfect the picture as a whole.

Note: The line draft needs to clearly express the overall dynamics of the characters, and the layout of the whole screen. First think about the general effect that you hope to present in your mind. In the hook line stage, it is necessary to properly position the details, and use some methods to make the whole appear smooth and keep shape.

In the light-colored marker stage, it is necessary to boldly paint the face and clothing, and it is possible to use the understanding of the human body structure for coloring. The transition color needs to be bold and fine-grained according to the turning position. The heavy color stage is the stage of grounding the picture, so that the whole picture does not appear too floating, and it is reasonable to boldly use the heavy color, embody the level and make the picture have space. The adjustment phase is also to let the painting reflect personal traits, and it needs to be adjusted according to your own ideas. The background also needs to be properly planned to be colored by selecting the appropriate color for the color of the subject.

案例一 波西米亚风格

波西米亚风格的服装主要特征是自由、随意，因此在画的过程中要控制笔触的节奏，不能过于僵硬，尤其是要把头发的动感表现出来。用马克笔作画的时候，要注意层次的体现，先画阴影，然后逐层用浅色马克笔画主面料部分，最后用白色高光笔勾高光线条（图5-1）。

Example 1 Bohemian Style

Bohemian Style is characterized by freedom and casual. Therefore, in the process of painting, the rhythm of the brush stroke should be controlled. It should not be too stiff, especially to express the dynamics of the hair. When painting with a marker, pay attention to the embodiment of the layer, first draw the shadow, then use the light-colored marker to draw the main fabric part layer by layer, and finally use the white color pen to highlight the line (Figure 5-1).

图5-1

画出 20 世纪服装风格 Drawing the Fashion Styles in 20th Century

用铅笔绘画线稿，然后用极细钢笔勾出外轮廓及细节图案；平铺出皮肤的颜色，涂出阴影，画好五官。给衣服上色，先用最浅的颜色平涂底色，然后逐层深入，最后画图案部分。色彩全部上完，再用粗钢笔给阴影部分断断续续地勾线

Draw shape with a pencil, then use a very thin pen to sculpt the outline and details; flatten out the color of the skin, apply a shadow, and draw the facial features. To color the clothes, first apply the base color with the lightest color, then darken it layer by layer, and finally draw the pattern. After all the colors are finished, use a thick pen to intermittently hook the shadows

第五章 服装风格的时装画表现及欣赏
Performance and Appreciation of Fashion Drawing

初定线稿——大体勾线——底色预铺——过渡色明确形状——重色压深——面部服装细节描绘——总体进行调整
Initial line drawing-general hook line–background color pre-laying-transitional color clear shape-heavy color depth-facial clothing detailing-overall adjustment

图 5-1

画出 20 世纪服装风格 Drawing the Fashion Styles in 20th Century

对面部和服装进行大量刻画，中间部分可适当弱化，形成强弱强弱的节奏变化。适当地加入一些自己想营造的画面效果，让整幅画具有趣味性

Depicts a large number of faces and garments, and the middle part can be appropriately weakened to form strong and weak rhythm changes. Appropriately add some effects you want to create, making the whole picture interesting

第五章 服装风格的时装画表现及欣赏
Performance and Appreciation of Fashion Drawing

因为服装的自身特色，需要突出服装的趣味性，着重刻画，选择大量的配色进行丰富，白色高光笔在这张画稿中起到了点睛的作用
Because of the unique characteristics of the clothing, it is necessary to highlight the fun of the clothing and focus on the portrayal. One could choose a large number of matched color to enrich painting, white highlight markers play a finishing role in this painting

图 5-1

画出 20 世纪服装风格 Drawing the Fashion Styles in 20th Century

初步定线稿对于动态的把握很重要，舒服的定位和流畅的动态可以为后期画面的深入打下坚实的基础
The preliminary alignment is very important for the dynamic grasp. The comfortable positioning and smooth dynamic can lay a solid foundation for the later stage of the picture

第五章　服装风格的时装画表现及欣赏
Performance and Appreciation of Fashion Drawing

戴银饰品是波西米亚风格的主要特点。先用细钢笔勾勒出饰品的形状，然后用浅灰色马克笔平涂出基本色；再用黑色彩铅笔画出阴影感

Silver jewelry is the main feature of the Bohemian Style. First use a thin pen to outline the shape of the jewelry, then use a light gray marker to flatly paint the basic color; then use a black pencil to draw a shadow

图 5-1　波西米亚风格
Figure 5-1 Bohemian Style

案例二 波普风格

画波普风格的时装画的关键点在于图案的画法，用平涂的方法画出服装上的图案，手法写实，色彩鲜艳（图5-2）。

Example 2 Pop Style

Pop Style fashion paintings are mainly about the drawing of patterns, using a flat-painting method to draw the patterns on the garments, and the techniques are realistic and paintings are colorful (Figure 5-2).

1960年代的波普风格的最大特点就是图案鲜明，色彩明快，在画的时候尽量如实画出图案，以平铺的方式涂出色块

The biggest feature of the Pop Style in the 1960s is that the patterns are bright and the colors are bright. When painting, try to draw the patterns as realistically as possible, and spread the blocks in a tiled way

第五章 服装风格的时装画表现及欣赏
Performance and Appreciation of Fashion Drawing

用卡通的画法表达1960年代初的小女孩风格
Cartoon drawing expresses Little Girlish Style of the early 1960s

图 5-2

画出 20 世纪服装风格 Drawing the Fashion Styles in 20th Century

创造性的画法有时候可以呈现不一样的体验
Creative idea sometimes can bring a different experience

第五章　服装风格的时装画表现及欣赏
Performance and Appreciation of Fashion Drawing

用细的勾线笔画出边缘线和图案部分，褶皱的部分要有抑扬顿挫的感觉；用彩色马克笔平涂图案部分；用一种颜色的马克笔画出皮肤色，再用深一度的马克笔画出阴影部分；用彩色画眼影。最后用粗的勾线笔在阴影面勾勒出阴影感
Draw the edge line and pattern part with a thin hook line pen, and feel the sensation in the fold part; flatten the pattern part with a colored marker; draw the skin color with one color marker, then use deeper markers shaded; the eye shadows are painted in color. Finally, a thick hook line pen is used to outline the shadow on the shadow surface

图 5-2　波普风格
Figure 5-2　Pop Style

案例三 解构风格

Example 3 Deconstruction Style

解构风格的服装最大特点是服装的解构是打破重组的,因此在画的时候要表现出夸张的感觉(图5-3)。

The most important feature of Deconstructed Style fashion is that the garments are broken and reorganized, so it is necessary to express an exaggerated feeling when painting (Figure 5-3).

用黑色的勾线笔勾勒出结构线和装饰线,然后用不同层次的灰色马克笔逐层上色

Use a black hook line pen to outline the structural lines and decorative lines, and then use different levels of gray markers to color the layers

第五章　服装风格的时装画表现及欣赏
Performance and Appreciation of Fashion Drawing

使用不同灰度的马克笔可以体现出层次感
Different grayscale markers reflect the layering

图 5-3

画出 20 世纪服装风格 Drawing the Fashion Styles in 20th Century

用白色勾线笔画出毛衣的高光效果
White hook line draws the highlights of the sweater

第五章　服装风格的时装画表现及欣赏
Performance and Appreciation of Fashion Drawing

为了体现头饰上的塑料质感，可以用浅灰色先平铺底色，阴影的地方用深一点的灰色和蓝色表现，再加一点粉紫色，这样就有了塑料材质的反光效果

In order to reflect the plastic texture on the headgear, you can use the light gray to flatten the background color, then the darker part of the gray and blue, and add a little pink and purple, so that there is a reflective effect of the plastic material

图 5-3

137

画出 20 世纪服装风格 Drawing the Fashion Styles in 20th Century

用简单有力的线条画头发的蓬松感
Draw the fluffy feel of the hair with simple and powerful lines

图 5-3 解构风格
Figure 5-3 Deconstruction Style

案例四　雅皮风格

Example 4 Yuppie Style

1980年代的雅皮风格服装的主要特点是夸张的肩部，因此在画的时候要将上下装形成鲜明的对比。另外，发型的画法要凸显出职业女性的干练特征（图5-4）。

The main feature of the Yuppie Style in 1980s was the exaggerated shoulders, so the top and bottom of the paintings should be in sharp contrast. In addition, the style of hair style should highlight the skill of professional women (Figure 5-4).

不同深浅的棕色和黄色来画卷发
Draw the curved long hair with different brown and yellow marker

图5-4

波点的画法：先用不同深浅的灰色画出服装的阴影，然后用灰色马克笔画出不同大小的点子

The way to draw dots: draw the dark part of the garments with different gray markers, then draw the dots with a gray marker

第五章 服装风格的时装画表现及欣赏
Performance and Appreciation of Fashion Drawing

边缘勾线一定要注意流畅性和虚实的变化，这样才能更好地表达时装画的动感
Edge line must be paid attention to the fluency and virtual and real changes, in order to better express the dynamics of fashion painting

图 5-4

画出 20 世纪服装风格 Drawing the Fashion Styles in 20th Century

白色勾线笔体现头发的高光
White hook line pen highlights the highlights the hair

第五章　服装风格的时装画表现及欣赏
Performance and Appreciation of Fashion Drawing

用不同蓝色的马克笔来体现层次感
Different blue markers to reflect the sense of hierarchy

图 5-4　雅皮风格
Figure 5-4 Yuppie Style

案例五　1920年代初期服装风格

20世纪初，在设计运动发展过程中，装饰运动紧随着新艺术运动之后发展，两者在风格上都强调装饰语言的运用。从形式上来看，装饰运动是新艺术运动的衍生和发展，他们之间有着共通之处。装饰运动看到了现代化和工业化发展的必然性，采用大量的装饰理念打破原有的机械形式。装饰艺术运动对时装产生的影响主要表现为：直线型、简洁的管子型和球型频频出现，这种风格既是时装的风格，也是与当时的建筑设计、产品设计、家居设计和平面设计的观念同出一辙。线描、平涂、装饰元素等都在画画中被格外强调。

20世纪初，以*Vogue*为代表的众多时尚刊物创办且发展兴盛起来，众多杂志主编，与各方人士保持着密切联系，关注当时正在迅猛发展着的现代艺术，鼓励各种新的时尚观念的表达，并在杂志中为一批年轻的艺术家们提供了时装化创作和表现的空间，并聘请了一大批生活在不同国家的艺术家为其创作杂志封面或插图（图5-5）。

Example 5　Fashion Styles in the Early 1920s

At the beginning of the 20th century, during the development of the design movement, the Art Deco followed the Art Nouveau movement, both of which emphasized the use of decorative language in style. Formally, the decorative movement was the birth and development of the Art Nouveau movement, and they had something in common. The decorative movement saw the inevitability of modernization and industrialization, and used a large number of decorative ideas to break the original mechanical form. The influence of the Art Deco movement on fashion was mainly manifested by the linear, simple tube type and the spherical appearance. This style was both a fashion style and the same concept of architectural design, product design, home design and graphic design. Line drawing, flat coating, decorative elements, etc. were all emphasized in painting.

At the beginning of the 20th century, many fashion publications represented by *Vogue* were founded and flourished. Many magazine editors kept close contact with all parties and paid attention to the modern art that was developing rapidly at that time, encouraging the expression of various new fashion concepts. In the magazine, he provided a space for fashion creation and performance for a group of young artists, and hired a large number of artists living in different countries to create magazine covers or illustrations (Figure 5-5).

第五章　服装风格的时装画表现及欣赏
Performance and Appreciation of Fashion Drawing

帽子是20世纪初期时装画中出现的一个非常重要的装饰元素
Hat is a very important decorative element of fashion painting in the early 20th century

图 5-5

145

画出 20 世纪服装风格 Drawing the Fashion Styles in 20th Century

直线型的造型
Straight shape

第五章　服装风格的时装画表现及欣赏
Performance and Appreciation of Fashion Drawing

装饰扣的画法
Decorative buckle painting

图 5-5

画出 20 世纪服装风格 Drawing the Fashion Styles in 20th Century

20世纪初的时装画很多是平涂画法
Many of the fashion paintings of the early 20th century were flat painting

第五章 服装风格的时装画表现及欣赏
Performance and Appreciation of Fashion Drawing

平涂画法
Flat painting

图 5-5　1920 年代初期服装风格
Figure 5-5　Fashion Styles in the Early 1920s

本章小结 Conclusions

1. 本章主要针对不同风格的服装讲解时装绘画的技巧和注意事项。
   This chapter aimed at teaching the drawing techniques.
2. 用马克笔画时装画时，不同层次的颜色叠加很重要。
   Dealing with the layers is very important in fashion drawing with markers.
3. 最后的勾线起着画龙点睛的作用，线条要有抑扬顿挫的感觉。
   Drawing the contour line is important, and the line should be drawn off and on.

思考题 Thinking questions

1. 在时装绘画中，马克笔和水彩画法的区别在哪里？
   What is the difference between marker and color painting in fashion drawing?
2. 用马克笔绘画时，如何表达层次的概念？
   How to draw the layers with markers?

参考文献
REFERENCE

［1］张玲. 图解服装概论[M]. 北京: 中国纺织出版社, 2005.
［2］袁春然. 时装时光. 袁春然马克笔图绘[M]. 北京: 人民邮电出版社, 2017.
［3］王受之. 世界时装史[M]. 北京: 中国青年出版社, 2002.